Qubuild: A Guided Approach to Asking Better Scientific Questions in Primary Schools

by Professor Lynne Bianchi and Christina Whittaker

Published December 2023 by Manchester University Press
ISBN: 9781526180070

© 2023 The University of Manchester

Contents

	page
About this book	5
What is the QuBuild concept?	9
Creating a classroom that is question-asking ready	10
Introducing the QuBuild Process	12
Identifying success with the QuBuild Process	14
How does the QuBuild Process relate to what scientists do?	17
The discipline of science – what scientists do	18
Recognising a scientific question	19
Providing a shared vocabulary for question talk	20
How does the QuBuild Process impact children's learning in science?	23
Improving the full enquiry cycle through using the QuBuild Process	24
Progression in science disciplinary knowledge using the QuBuild Process	25
How to embed the QuBuild Process into classroom practice	29
Getting going with the QuBuild Process	30
Structuring learning to get better at scientific question-asking	31
Classroom resources and teacher guidance support	32
Step 1: Question-producing: children generating questions	32
Step 2: Question-handling: children working with questions	48
Step 3: Question-improving: children refining and choosing scientific questions	64
How does the QuBuild Process affect professional learning?	85
Teacher talk examples of the QuBuild Process in context	88
References and recommended reading	94
Acknowledgements	95
About the Authors	96

Questions and questioning underpin the foundational habits of mind of scientists.

Çalik et al., 2012

About this book

This book brings a new classroom approach for teachers who rightly recognise that teaching the explicit knowledge of scientific question-asking is not simple!

What better way to introduce a book about asking questions than to ask some questions of those in primary classrooms?

We asked:

When do teachers ask questions in primary classrooms?

Teachers said:
Every classroom, every day, every teacher asks questions.

Children said:
All teachers ask lots of questions!

When do children ask questions in primary classrooms?

Teachers said:
Every classroom, most days, most children ask questions.

Children said:
Sometimes there is a chance to ask questions that matter to me, but sometimes my questions go unheard.

When do teachers ask scientific questions in primary classrooms?

Teachers said:
Some classrooms, sometimes, some teachers depending on their confidence and subject specialism.

Children said:
What is the difference between a scientific question and a normal question?

When do teachers teach children to ask scientific questions in primary classrooms?

Teachers said:
Some classrooms, with some teachers, might talk about the features of a scientific question before an enquiry or practical activity.

Children said:
Teachers tell some children that they have asked a good question but we're not sure why.

When do children ask scientific questions in primary classrooms?

Teachers said:
Every science lesson most children ask various and many questions.

Children said:
The teachers are very good and helpful at giving the answer to the questions asked.

> **A high-quality science education is rooted in an authentic understanding of what science is.**
>
> Ofsted 2021, Research Review Series: Science

What is the QuBuild concept?

Primary classrooms are busy places. During their busy days, teachers and children have multiple opportunities to ask and answer questions. Questions are asked every hour of every day. The reality is that there are never enough hours in a day. The time spent on questions needs to be thoughtfully planned for so that it forms a meaningful learning opportunity that results in high-quality outcomes. It is therefore vital to create a classroom culture that is 'scientific question-asking ready'.

Creating a classroom that is scientific question-asking ready

A classroom culture for scientific question-asking can be created by positively encouraging dialogue, collaboration, curiosity and critical thinking. A classroom that is ready to benefit fully from the QuBuild Process requires all four values to be explicitly highlighted and rewarded.

Developing a scientific question-asking classroom culture is best supported when:

Dialogue is encouraged and active

Children's questions and thinking are heard, explored and reacted to. Discussion between children and teachers enables uncertainty in science learning concepts to be explored, and for children to think out loud. The link between talking, thinking and reflection on learning is valued. Getting better at scientific question-asking is an iterative process and should be shared and thought about together.

Collaboration between children is encouraged

Classroom routines expect group work that is inclusive of all children. Groups arriving at jointly agreed scientific questions is authentic to the way scientists work. Children working in small groups of two or three is ideal as it gives the opportunity to share their ideas, adapt them together and jointly own the outcome. This engenders cooperation and shared responsibility for the outcomes.

Curiosity is given time to flourish and be celebrated

Although time is always at a premium, classrooms which value better scientific questions will need to afford it. Good ideas, reflection and refinement need time. Practical experiences, reading, outdoor exploration, discussion and making links to prior learning are all helpful ways to stimulate curiosity. When children experience curiosity-rich classrooms, there is a greater chance for a wide range of interesting questions to be asked that can be developed into better scientific questions for enquiry.

Critical thinking is supported and expected

Children engage in making sense of evidence they collect, and build new ideas. Thinking critically, considering, summarising, evaluating and recommending are all higher order skills that need to be learnt through modelling and practice. Vital to getting better at asking scientific questions will be the classroom opportunities where teachers facilitate, demonstrate and guide the learning process.

Children's science learning is going well when...

Children are encouraged to use their own curiosity, scientific interests and questions

- Children use their own ideas to answer questions

- Children construct their own science questions and know how to answer them

- Children ask and investigate their own scientific questions

Bianchi, Whittaker & Poole, 2023

Introducing the QuBuild Process

This book gives teachers a new strategy for classrooms to improve how children can get better at asking scientific questions. This new teaching and learning strategy is referred to as the QuBuild Process. The QuBuild Process is structured in three steps.

The book guides teachers through each step to understand the rationale behind them. Resources enable them to use the process with children in the classroom.

The importance of the QuBuild Process

All of us who have worked with children in primary science recognise the essential experiences gained from practical investigations. Finding out about the world through enquiry, where children have autonomy to explore concepts and ideas using the scientific method is fundamental to primary science. What many often fail to recognise is the challenge that asking scientific questions, and asking good scientific questions, poses to even adults.

It certainly isn't simple!

Think back to the first time you tried to tie a shoelace. You probably didn't simply watch someone do it and repeat it. To master a skill requires understanding of the smaller simpler steps, trying them out and practising them repeatedly until they come together and what initially seemed complex is achieved. Now think about a more challenging skill, such as learning to play a new musical instrument. This time, as well as the small steps, with explanations and practice and by engaging with guidance and support, you can continually progress and get better and better at the skill.

Asking a scientific question is a challenging skill

Asking questions is a skill that we practise from an early age, and to get better at doing so requires repeated practice, development of the steps involved and the guidance to develop complexity in that skill. It is our responsibility as teachers to provide the essential experiences to enable children to learn, understand, practise and master the skill of scientific question-asking. Without this, we risk encouraging children to investigate the world around them, whilst failing to deepen their curiosity through better questioning. Question-rich classrooms can provide the support for children to grow and develop successfully as primary scientists – or indeed future scientists.

Making a difference through QuBuild

Through QuBuild children are taught the tactics and strategies to become successful in each of the QuBuild steps. Ultimately children will be able to use the QuBuild Process independently and confidently in new contexts for their questions about their new learning.

By 'building' better scientific questions children will gather evidence during science enquiry that better supports them to draw meaningful conclusions that enhance their understanding about the world around them. Consequently, not only do children get better at question-asking, but they get better at primary science learning too.

Producing questions is the essential first step of the QuBuild Process. To have questions that can be improved, requires that there are a collection of questions to start with. For this, teachers and children stimulate curiosity within the topic of study. Children work together to generate lots and lots of questions that they are curious about.

Once children have a collection of questions, it is important that they are engaged in analysing them. Handling questions is the essential second step of the QuBuild process. For this teachers enable vocabulary acquisition so that children can apply their knowledge of question types and question features to identify, group, sort and organise the range of questions that they have produced.

By selecting a few questions from those sorted in the previous step, children now critique and find ways to improve them. Question improving and choosing is the final step of the QuBuild Process. Children improve their scientific questions and prioritise those which they can take forward into a scientific enquiry.

When brought together we have 'The QuBuild Process'.

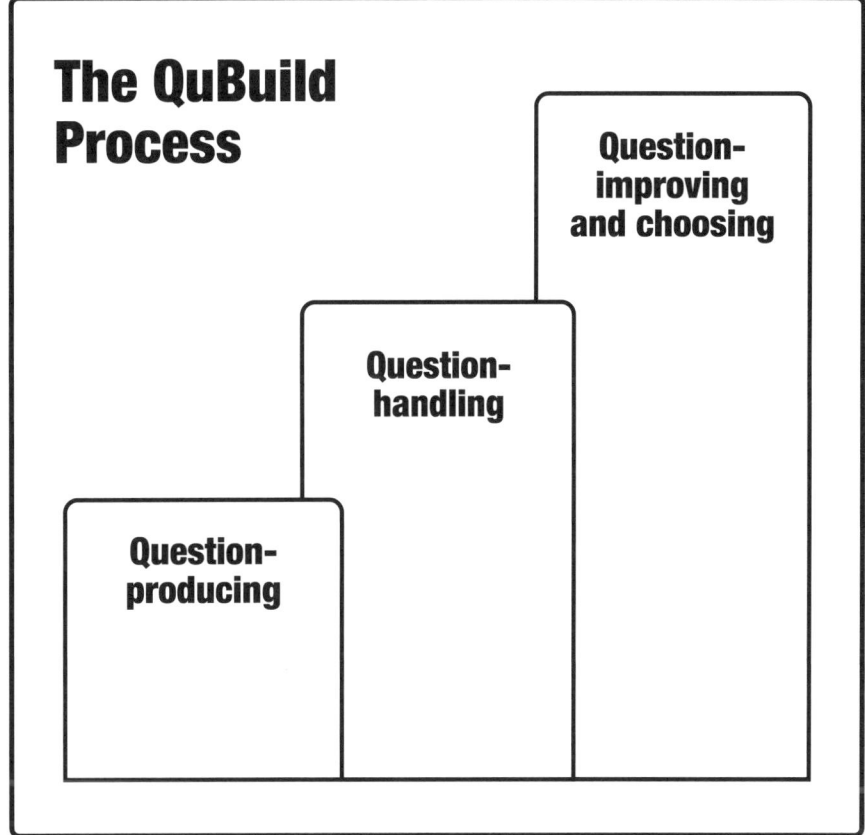

The authors acknowledge the influence that the Question Formulation Technique (QFT) had to their thinking in the development of the QuBuild Process. QFT was created by The Right Question Institute and detailed in Rothstein, D. & Santana, L. (2015) Make Just One Change: Teach Students to Ask Their Own Questions. Harvard Press.

Identifying success within the QuBuild Process

	Learning Outcomes	Success is when...
Step 1 Question-producing	To be able to produce more questions of a wider variety.	Children understand that the first question is not the only question.
Step 2 Question-handling	To be able to compare and contrast different questions providing reasons for similarities and differences.	Children understand the features and characteristics of different questions.
Step 3 Question-improving and choosing	To be able to recognise that some questions are better than others.	Children understand the relationship between the quality of the question and evidence.

Evidence is fundamental to the science enquiry process.

Identifying success from the QuBuild Process

	Learning Outcomes	Success is when…
The QuBuild Process in full	To know that a well-designed question improves the evidence gathered and conclusions drawn.	Children acquire more knowledge and deeper understanding through their enquiries as the evidence focuses on authentic new learning.

What is QuBuild?

A focused approach to plan for better science learning for primary children.

When they are following their own noses, learning what they are curious about, children go faster, cover more territory than we would ever think of trying to mark out for them, or make them cover.

Holt, 1971

How does the QuBuild Process relate to what scientists do?

When promoting the desire for children to work as scientists it is useful to consider what is special about being a scientist. Skills taught through the QuBuild Process relate positively to work published by the University of Berkeley intended to improve a shared understanding of what it means to work scientifically.

The discipline of science – what scientists do

Science is a creative human endeavour which builds new knowledge to explain natural phenomena. It is an empirically-based process derived from observation of the natural world. Using the QuBuild Process, children will be working as scientists by producing scientific questions that they are curious about.

Science is a rigorous discipline where it is important to know how the evidence was collected and whether it can be trusted. Using the QuBuild Process, children will be working as scientists by handling and sorting scientific questions based on suitability, credibility and relevance.

Scientific knowledge is tentative and subject to change based on new evidence or new interpretation of existing evidence. Using the QuBuild Process, children will be working as scientists by improving and choosing scientific questions for gathering quality evidence.

Science is described as universal, carried out in all cultures at all ages, creating a diverse scientific global community. Children in QuBuild classrooms engage in collaborative groups and develop the skills of critical thinking and peer-review.

Question-rich classrooms provide the support for children to grow and develop successfully as primary scientists – or indeed future scientists.

Recognising a scientific question

Asking questions is something that children and adults do every day. So many questions are asked about so many things. Not all questions are the same type. There are different types of questions that seek to gain different types of answers, for instance those such as 'What time is it?', or 'Can I go to the toilet, please?' and 'Where can I find the equipment?' etc. In science, the questions that teachers expect to hear are specific and described as scientific questions.

For children to get better at asking *scientific* questions, we need to understand and be clear about what makes a **scientific question** unique and different to all sorts of other questions.

Science involves seeking answers to questions through observations and measurement, collecting, analysing, and reviewing data. By gathering evidence, children can make sense of things that are uncertain and that they are not sure about. Some evidence may actually help to change and challenge what might have previously been considered as certain. Evidence helps to build conceptual understanding of scientific ideas and principles.

What is a scientific question?

A scientific question leads to evidence being gathered and conclusions being drawn, so that more knowledge and understanding is developed.

Evidence is fundamental to the science enquiry process. Understanding the importance of evidence when planning an enquiry is vital if we are to maximise the learning impact in our science classrooms. **Asking better scientific questions is at the heart of better scientific enquiry.**

The QuBuild Process empowers children to talk about, know about and practise developing their scientific questions with each other. They will become aware of how fundamental evidence is to scientific enquiry and be better able to identify scientific questions that challenge, broaden or deepen their knowledge of the world around them.

Providing a shared vocabulary for question talk

There are many different types of questions used in the world. In this book, we focus on six question types. These provide specific vocabulary for children to use when talking about questions. This enables them to think about and describe questions in the primary science classroom. As it is valuable to develop consistent language to support meaningful learning we have selected six types of questions that are relevant to the QuBuild Process.

Question types in action

Question type	Description of the question type	Examples of questions
Closed questions	Require a 'Yes-No' answer. Open questions are all other questions that do not result in a 'Yes-No' answer	Is copper magnetic? Do all living organisms reproduce? Was Mae Jemson the first woman in Space?
Opinion questions	Cannot be answered by scientific enquiry, they are about what people think, feel or believe	Are woodlice scary? Are rainbows beautiful? How do you feel about the dark?
Fact-find questions	Can be answered using an information search, agreed definition, generally descriptive	What is a magnet? What comes out of an erupting volcano? Who first explained gravity? What are the names of the planets? What are the names of different types of rocks?
Empirical questions	Are based on observing a change or effect, not a theory	How far will my paper plane travel? How fast does ice melt? Which paper towel soaks up the most water? What do the moulds that grow on bread look like after three weeks?
Cause-effect questions	Evaluate the effect of one thing on another. Includes at least two variables	If we stretch the elastic-band to different lengths, what will happen to the pitch of sound when we pluck it? How does the shape of the glider's tail affect the distance that it will travel? How much does your heart rate change when you do different exercises? When I change the size of the spinner's wings, what happens to the time it takes to fall?
Trust questions	Evaluate the certainty of information and its trustworthiness	How sure are we that plastics are harming the planet? How can we be certain that medicines work? What proof is there that exercise improves our mood? Can we trust adverts that say that a particular product works? How certain are we that eating 5 fruits and vegetables each day makes us healthier?

QuBuild Question Keywords

Question type	Description of the question type
Closed questions	Require a 'Yes-No' answer. Open questions are all other questions that do not result in a 'Yes-No' answer
Opinion questions	Cannot be answered by scientific enquiry, they are about what people think, feel or believe
Fact-find questions	Can be answered using an information search, agreed definition, generally descriptive
Empirical questions	Are based on observing a change or effect, not a theory
Cause-effect questions	Evaluate the effect of one thing on another. Includes at least two variables
Trust questions	Evaluate the certainty of information and its trustworthiness

The right kind of scientific question, for the right purpose, at the right time, for the right reason.

Bianchi & Whittaker, 2023

How does the QuBuild Process impact children's learning in science?

Most schools use a continuous school development cycle in order to improve children's learning experiences and outcomes. In science teaching and learning this can result in changing from doing practical activities because they are 'fun' to having specific areas of focus mapped to the enquiry cycle. The QuBuild Process can result in further improving working scientifically in a rigorous and considerate way.

Improving the full science enquiry cycle using the QuBuild Process

Teaching children to get better at asking scientific questions can have a positive benefit on the children's experience when using the enquiry cycle. Asking a scientific question is often the first stage in an enquiry and the quality of this question has a direct impact on the way children plan, how they observe and measure, record, interpret and report and evaluate.

The following diagram provides an illustration of the potential impact the QuBuild Process can have on the enquiry cycle. By seeing the value of question-asking to all the stages within enquiry it becomes possible for teachers to consider what currently is working well and what might be a focus for improvement.

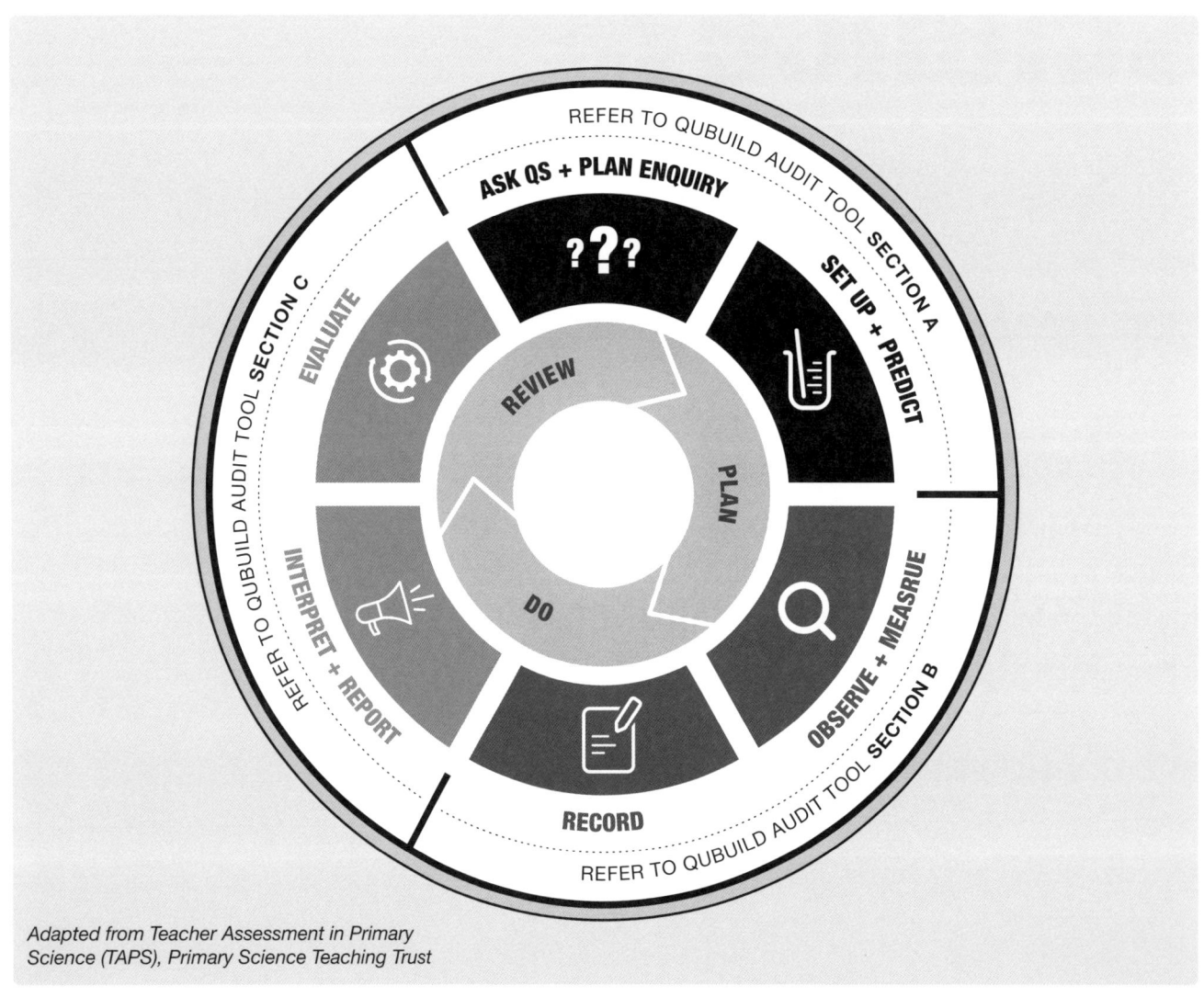

Adapted from Teacher Assessment in Primary Science (TAPS), Primary Science Teaching Trust

Progression in science disciplinary knowledge using the QuBuild Process

The QuBuild Process is relevant to all age groups. Developing and honing the skill of active scientific question-asking means that the learner can appreciate that the first question they ask may not necessarily be the best one.

The way a question is formed influences its purpose and how evidence or results are gathered. The more honed the question, the more likely that the evidence that is gathered will be trustworthy.

By planning the QuBuild Process into the curriculum, children can progress in their development by becoming:

- **increasingly independent** in asking scientific questions using and selecting appropriate strategies and prompts without teacher intervention.
- **increasingly systematic** in asking questions that are appropriate and fit for the enquiry approach identified.

By planning the QuBuild Process repeatedly into the curriculum, children can practise and progress in asking questions so that their enquiries are:

- **increasingly precise** meaning that the better questions will reduce or allow for errors in the observations or data to be considered.
- **increasingly valid** meaning that the better questions will identify relevant and irrelevant variables when seeking patterns.

Over time, children will ask better scientific questions which are more justified and fit-for-purpose. The better scientific questions result in more valid conclusions. Children will become better at taking responsibility for enhancing the quality of the evidence they gather and they will become better at linking their own findings to create meaningful new learning.

> The better the scientific question, the more likely it is to lead to quality outcomes during the enquiry.

Reflecting on your own practice

The QuBuild Audit Tool (see the following table) shows the extremes of children's behaviours related to science enquiry in the classroom. Use this to engage teachers in self and peer-reflection, rating the children's knowledge of different enquiry skills between two extreme statements. This will challenge assumptions of current practice and by working with your colleagues you will be able to decide on a shared vision of what good practice in enquiry looks like in your school and you'll be ready to begin the QuBuild Process.

The QuBuild Audit Tool can also be useful to evaluate whether the QuBuild Process has had an impact. The recommendation is to revisit the auditing activity systematically and regularly over time.

Copy the audit tool, reflect and review the behaviours you most regularly see your children using. Tick where on the continuum between 'From' and 'To' you consider the children to be now. What does this reveal to you about their knowledge and experience of science enquiry? Where would you wish them to be?

Teacher QuBuild Audit Tool

A. *Children's knowledge of:* Asking questions, planning and setting up enquiry

From	⟶	To
Children take the question from the teacher, or stick with the first question that seems sensible-enough.	☐ ☐ ☐ ☐ ☐ ☐	Children know that the first question isn't always the best question, and that they need to consider how it could be developed/improved for themselves.
Children having a limited understanding of what scientific questions are, often seeing this as cause-effect only.	☐ ☐ ☐ ☐ ☐ ☐	Children have a deeper understanding and recognition of what makes a scientific question and know that there are different types of questions.
Children are over reliant on teacher talk and direction, they lack autonomy and independence with the question to explore pre-set by the teacher.	☐ ☐ ☐ ☐ ☐ ☐	Children proactively ask questions in response to a context that they have been inspired by.
Children are engaged in prescriptive practical work.	☐ ☐ ☐ ☐ ☐ ☐	Children see themselves as being reliant on each other to collaboratively make decisions about what to investigate, what not to investigate and why, and then to bring the teacher into that process.

B. *Children's knowledge of:* Observing, measuring and recording

From	→	To
Children work in pairs or small groups to do an investigation in a logistical and mechanistic way.	☐ ☐ ☐ ☐ ☐ ☐	Children work in pairs or small groups to discuss and debate – cognitively and creatively.
Children lack perseverance because they're not intrinsically motivated to find out.	☐ ☐ ☐ ☐ ☐ ☐	Children are inspired and engaged in the enquiry process, showing resilience because they genuinely want to find out.
Children all complete a standard pre-designed format to enter their results.	☐ ☐ ☐ ☐ ☐ ☐	Children appreciate that communicating science evidence is considered with its appropriateness to aid peer review and scrutiny of new findings.

C. *Children's knowledge of:* Interpreting, reporting and evaluating

From	→	To
Children are not always sure how what they've found out helps them, often because an investigation hasn't been initiated by a scientific question.	☐ ☐ ☐ ☐ ☐ ☐	Children identify and articulate how their results/evidence supports them in answering their scientific question.
Children find it difficult to draw a conclusion or their conclusion is broad and lacks clarity.	☐ ☐ ☐ ☐ ☐ ☐	Children reflect on evidence to draw conclusions that directly relate to the question.
Children can't connect their science learning to what they know and understand already.	☐ ☐ ☐ ☐ ☐ ☐	Children can connect their learning to what they know and understand already, making it more meaningful.
Children worry about getting science 'right', getting the 'right' results.	☐ ☐ ☐ ☐ ☐ ☐	Children are willing and comfortable to work on a question they're not familiar with and that they genuinely don't know the answer to, recognising that science develops over time.

Fostering students' learning goals that produce more and better questioning remains one of the most important challenges in the teaching process.

Costa et al, 2000

How to embed the QuBuild Process in classroom practice

Motivated leaders of science are continuously developing approaches to critique the effectiveness of the way they teach children to ask better questions. Introducing change that can be sustained and embedded needs to be considered and planned for.

Getting going with the QuBuild Process

For many teachers, this book will be the first time they have focused so intently on one specific scientific skill. The first read through will stimulate professional awareness of the complexity of asking good questions and enable an appreciation of assumptions made that because many questions are heard in the classroom all is fine. Having raised the issue of teaching scientific questions as hard, teachers will be pleased to find that there are teaching approaches and suggestions to help address the issue.

Sharing and talking about the ideas presented in this book with an enthusiastic colleague or within school clusters or networks is an ideal approach to effective professional development. It is important to appreciate that getting to grips with the concepts presented may take two or three active reads.

For science teachers who want to improve children's learning through the ideas in this book, our recommendation is to trial them in small chunks, to drip feed the stages, to build confidence and to reflect on the challenges of trying something new. There is no need to take all the ideas and try them all out immediately as this risks cognitive overload for both the teacher and the children.

By developing personal experience and expertise in using the QuBuild Process, subject leaders can apply their learning from their classroom to a whole school approach. This will require support and interest from colleagues through staff meetings, coaching and regular reflective dialogue.

The more honed the question, the more likely that the evidence that is gathered will be trustworthy.

Structuring learning to help children get better at scientific question-asking

The QuBuild Process involves three steps. The resources for each step, 'question-producing', 'question-handling' and 'question-improving and choosing', have been designed through a learning pathway that starts with teacher input and worked examples through guided support towards independence through scaffolded prompts. Each step has the same structure for the resources and guidance.

Introductory Activities

Introductory Activities to **activate children's learning**. Children will be developing their knowledge or acquiring new thinking explicit to the step by using the introductory activities. These activities are not intended to be used repeatedly and are offered to allow teachers to introduce key vocabulary and new ideas.

Learning Tools

Learning Tools to support children to **begin to practise**. Children will be able to apply new learning to their own questions by using the learning tools. The skills of question-producing, handling and improving will be used increasingly independently as children have a technique upon which to lean for guidance. These tools are intended to be used repeatedly regardless of lesson context.

Visual Prompts

Visual Prompts as **icons** to support children to remember and recall what each of the different steps involves. These prompts or icons are intended to be a short, quick and easy structure to refer to during the flow of the lesson, as and when appropriate.

Step 1
Question-producing: children generating questions

Why do we need this step?

For children to know that creating an effective question takes time and is a big part of the work scientists do.

What does this step involve?

Children generating and harvesting their own questions. All children being included and having their questions valued.

The essential aspects of this step are when children engage fully by:

- **coming up with questions from a stimulus provided by their teacher**
- **having no restriction on how many questions they come up with**
- **producing a variety of questions**
- **pooling their questions together.**

Teacher Notes

Although time is always at a premium in our classrooms it is essential that teachers value the openness of this step. The danger is limiting children's questions at this stage will limit the rest of the QuBuild Process.

What is question-producing about?

This step results in children having the chance to think freely without concern of the quality of their initial questions. Importantly, every child has a chance to have their question included and listened to. As a fast-paced activity it is likely that the only sound in the room will be that of children asking questions. They are encouraged not to think that their first question is the 'only' or 'best' question to ask.

By using this step, the expected impact will be:

From		To
Children asking a few questions or a similar sort		Children asking many and varied questions
Teachers intervening too soon to amend or improve questions for the children		Teachers assisting children to gather as many questions as possible, reserving judgement and not intervening to improve questions in the first instance
A few questions being valued		All questions being valued
Children feeling that they have to get the question 'right' and 'as good as it can be' the first time round		Children knowing that questions will be changed and improved at a later stage
Children assuming the question is the best because it's theirs		Children having a sense of team-ness or group ownership of questions through the pooling of their questions with others

What knowledge do children need in order to improve their question-producing?

> Children benefit from knowing that scientists can produce questions by:
> - observing the world around them
> - imagining possibilities
> - looking for clues or patterns.
>
> Children benefit from knowing that there are many different scientists and different jobs that they do, which results in many types of questions being produced.

Classroom Resources provided to support Question-producing

Introductory Activities
- Is it a question?
- Observe-Imagine-Look

Learning Tools
- Question Spinner
- Question Frames
- Wonder Bubbles
- Question Teller

Visual Prompt
- Question-producing visual icon

Step 1: Question-producing

Introductory Activities

Is it a question?

What's it for?
Children learn to distinguish between a question and a statement. In the examples provided, the question mark has been removed so that children have the opportunity to think about the structure of the phrase to determine if it is a question or not. Children can practise where a question mark is appropriate or not.

How does it work?
Prepare ahead of the activity 'Question and Statement examples' cards as provided below. Children should select those cards from the set that represent a question, and separate them from those that represent statements. Children should recognise that the questions require a question mark. Teachers can extend the activity by asking the children to create additional cards that they can sort with peers. A further extension of the activity is to ask the children to adapt statements into questions and questions into statements.

What to expect when it is going well?
Children should be able to explain the difference between a question and a statement and articulate the choices that they make.

> **Teacher Notes**
>
> See page 36-37 for the 'Children's Question and Statement Examples' to copy for your classroom.

Statements make assertions. Being a statement, however, does not necessarily make the assertion true. Evidence is required to trust a statement.

Teacher Guidance (Correct Responses)

Questions	Statements
What colour flowers do pollinating insects prefer?	Eggshells and teeth are damaged by acid
How does the length of the carnation stem affect how long it takes for the food colouring to dye the petals?	Boys in our class have bigger heads than girls
How do the teeth of mice differ from the teeth of hamsters?	Teeth are made from enamel
Do all flowers have the same number of petals?	The most common number of petals on flowers in the school grounds is 5
What are all the different ways that seeds disperse?	Metals conduct electricity
Are the lifts in tall buildings safe?	The tallest building is in Dubai
How much faster does sugar dissolve in hot water than cold?	Older people have grey hair
How many planets are in the solar system?	The giant tortoise was found living in the Galapagos Islands
Why do people get grey or white hair as they get older?	Push and pulls are forces
What happened when Charles Darwin visited the Galapagos Islands?	The Earth moves around the sun
How does an egg shell change when it is left in cola?	The length of a shadow changes during the day
Do coloured objects make coloured shadows?	Seeds do not need light to germinate

Step 1: Question-producing

Children's Question and Statement examples

| What colour flowers do pollinating insects prefer | Eggshells and teeth are damaged by acid |

| How does the length of the carnation stem affect how long it takes for the food colouring to dye the petals | Boys in our class have bigger heads than girls |

| Teeth are made from enamel | Seeds do not need light to germinate |

| Do all flowers have the same number of petals | The most common number of petals on flowers in the school grounds is 5 |

| Metals conduct electricity | What are all the different ways that seeds disperse |

| Are the lifts in tall buildings safe | The tallest building is in Dubai |

How to embed the QuBuild Process in classroom practice

How much faster does sugar dissolve in hot water than cold	Older people have grey hair
The giant tortoise was found living in the Galapagos Islands	How many planets are in the solar system
Why do people get grey or white hair as they get older	Push and pulls are forces
The Earth moves around the sun	What happened when Charles Darwin visited the Galapagos Islands
How does an egg-shell change when it is left in cola	The length of a shadow changes during the day
Do coloured objects make coloured shadows	How do the teeth of mice differ from the teeth of hamsters

Step 1: Question-producing

Observe-Imagine-Look

What's it for?
Children learn about three approaches scientists might use to produce questions. They may:

- observe the world around them
- imagine possibilities
- look for clues or patterns.

How does it work?
There are three headers provided of the approaches some scientists use to produce questions and these are to be made easily visible in the classroom when doing this activity, e.g. as a list on the board, posters on walls or labels for sorting hoops.

HEADER 1:
Observing the world – scientists create questions by observing the world around them.

HEADER 2:
Imagining possibilities – scientists create questions by imagining possibilities.

HEADER 3:
Looking for clues from patterns – scientists create questions by looking for clues or patterns.

Ahead of the activity, prepare the 'Observe-Imagine-Look' cards from the resource so that each group of children can be provided with a set ready to sort. Ask the children to match the job description to the header of best fit. Accept any reasonable considerations as there can be multiple outcomes from this activity, debate between groups is to be actively encouraged.

What to expect when it is going well?
Children recognise that there are different ways of thinking scientifically to produce questions. Children recognise that they can also start to use the three approaches to create their own questions by thinking as scientists.

Teacher Notes (Correct Responses)

Observing the world	Imagining possibilities	Looking for clues from patterns
Geologist	Sound engineer	Nutritionist
Herpetologist	Textile engineer	Forensic scientist
Zoologist	Architect	Architect
Entomologist	Aerospace engineer	Conservationist

How to embed the QuBuild Process in classroom practice

Children's Observe-Imagine-Look Cards

| **Observing the world** | **Imagining possibilities** | **Looking for clues from patterns** |

Geologist
A scientist who observes the solid, liquid and gaseous matter in our world including the planets

Conservationist
A scientist who studies patterns in animal and plant behaviour and acts for the protection and preservation of the environment and wildlife

Entomologist
A scientist who observes insects in the world around us

Forensic scientist
A scientist who looks for clues and traces of physical evidence for use in courts of law

Textile engineer
An engineer who designs and creates fabric, and the processes, equipment and materials needed to create different fabrics

Herpetologist
A scientist who observes reptiles and amphibians

Architect
An engineer who is qualified to imagine, design buildings and to plan and supervise their construction

Zoologist
A scientist who observes the behaviour, physiology, classification and distribution of animals

Meteorologist
A scientist who collects and studies patterns in data from the atmosphere and oceans that make weather

Nutritionist
A scientist who looks into different foods and the impact of nutrition on different people

Aerospace engineer
An engineer whose work focuses on the creative design, development, testing and operation of aircraft and spacecraft

Sound engineer
An engineer who explores how high quality live or recorded sounds can be produced

Step 1: Question-producing

Learning Tools

Question Spinner

What's it for?
Children learn about question starter words (question stems) and use simple approaches to randomly select them to inspire their own question-asking.

How does it work?
Print and cut out the spinner template from the resource below. Place a paperclip over the centre of the circle. Then place the tip of a pencil through the paper clip onto the centre of the circle. Flick the paperclip to let it spin. Children should ask a question using the word that the paperclip lands on.

What to expect when it is going well?
Children ask a range of questions and many questions. Children begin to recognise the variety of question starter words they can use to ask questions about the world around them.

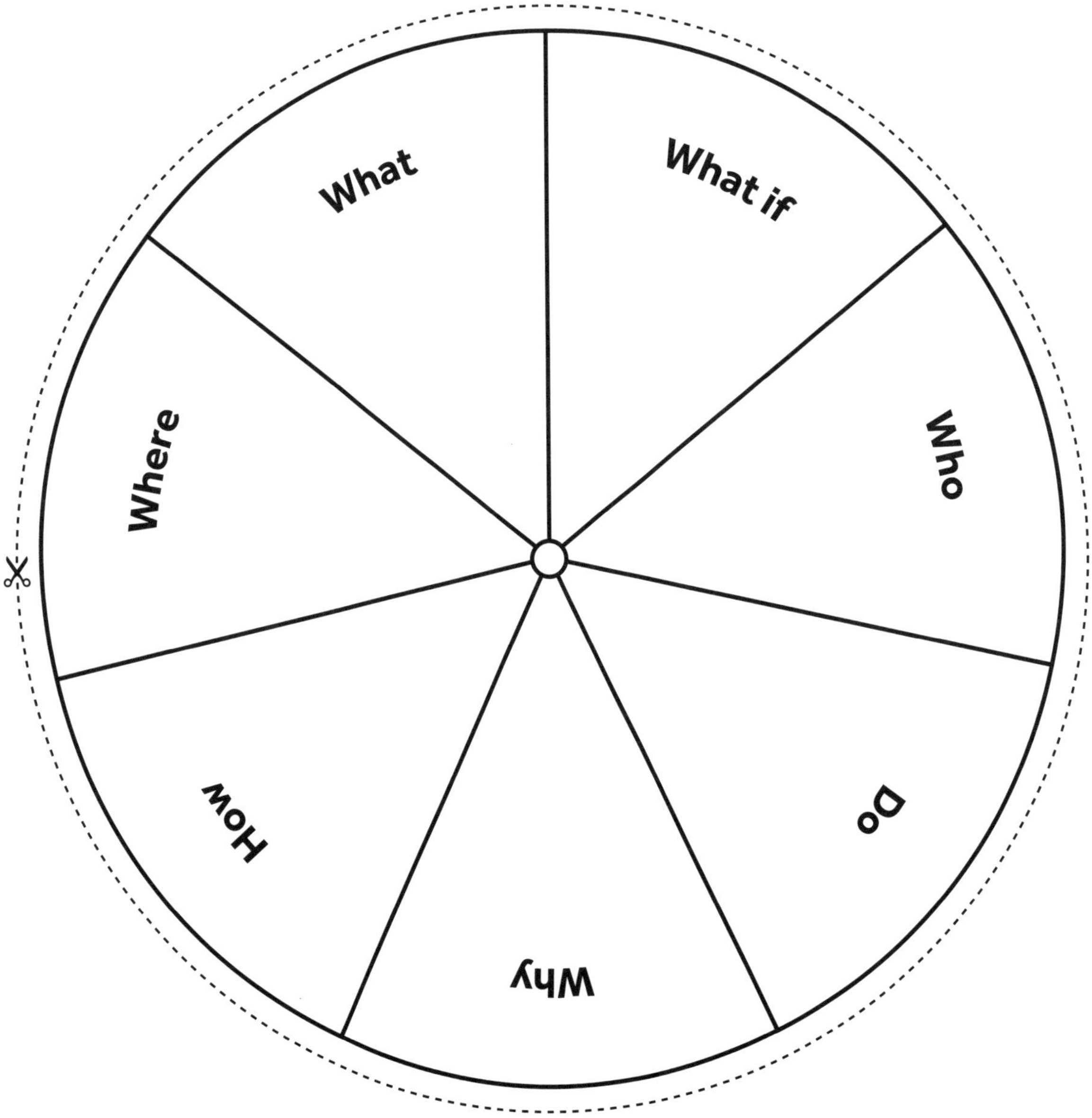

Children's Question Spinners

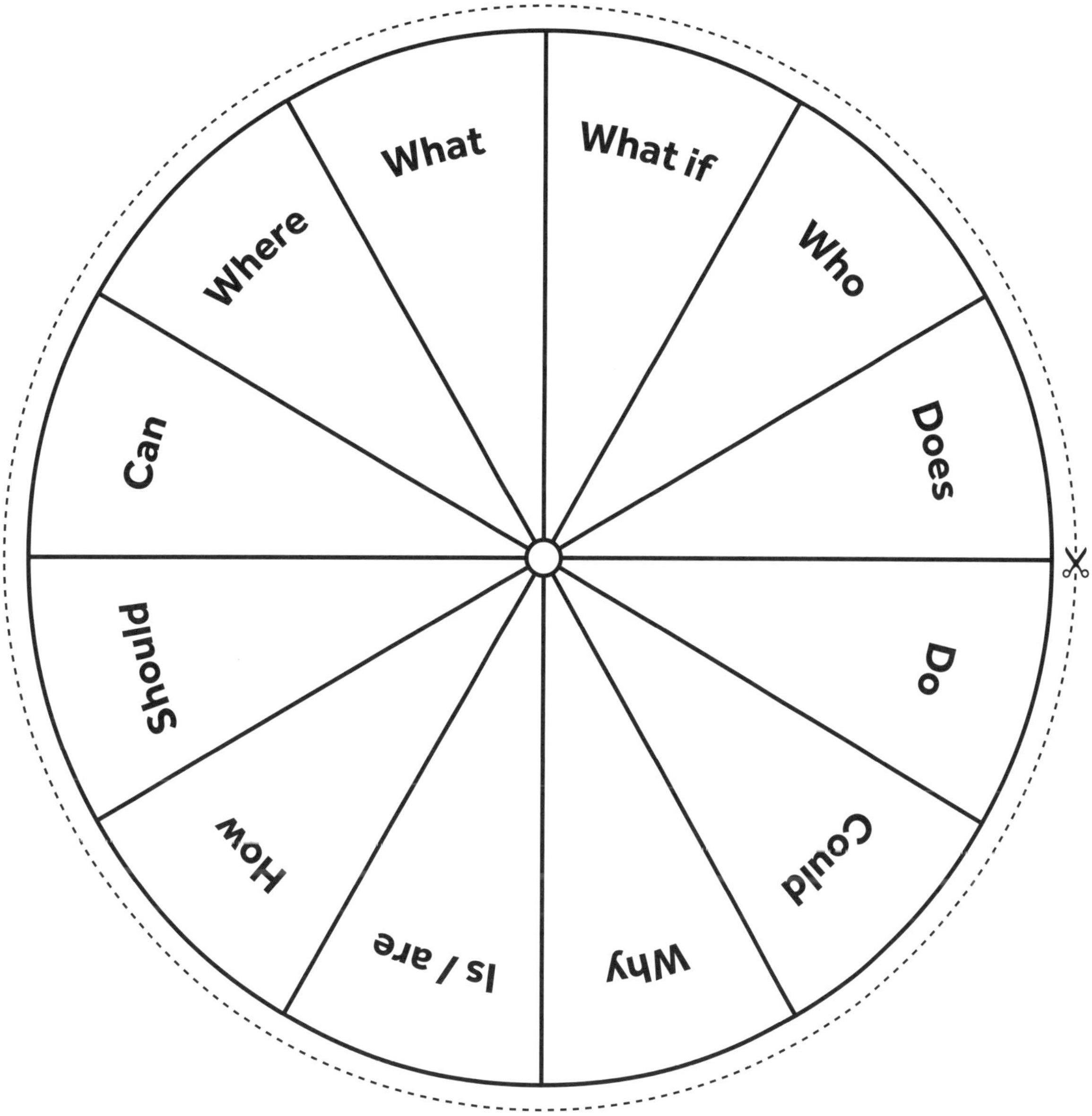

A circular spinner divided into 12 segments with the following question starters: What if, Who, Does, Do, Could, Why, Is/are, How, Should, Can, Where, What.

Teacher Notes

Select the 'Question Spinner' that best suits your children. You may develop the use of the spinner using the simpler one, and extend question starters using the second.

Step 1: Question-producing

Question Frame

What's it for?
Children produce questions that result from a close observation of the features and characteristics of an object they have chosen to study.

How does it work?
Use the suggested design to make a frame out of an old cardboard box or paper. Ask the children to place the frame over the object being observed, so that it appears in the centre. Teachers may select different objects depending on the topic or lesson theme or children might be free to find an object of their choice within the topic. Ask children to talk together about what they can see and describe it to each other. The questions are to be written onto sticky notes and placed to the side of the frames. The sticky notes provide a way to prioritise and select the question(s) the group wish to share with the class.

What to expect when it is going well?
Children produce many questions in response to what they observe closely. Their questions are focused on an object under investigation.

Wonder Bubbles

What's it for?
Children develop their knowledge of question starter words by adding the phrase 'I wonder…' before them. This approach further emphasises the 'possibility thinking' that scientists use to stimulate questions. By using the 'I wonder…' phrase, children are stimulated to be increasingly open-minded when question-asking.

How does it work?
Ahead of the activity prepare a class set of the 'Wonder Bubbles' cards. Encourage children to spend time looking or walking around the spaces near them. Encourage them to carefully and safely look, listen, hear, touch and smell things around them. Children spread the cards on a table-top and then choose as many as they wish to describe different 'wonderings' they have. Encourage them to find other people with similar 'wonderings' and to talk more about the questions they have.

What to expect when it is going well?
Children ask a range of questions that are imaginative as they are inspired to think about the world around them. They use the wonder phrases to vary their ideas and phrase them as questions.

Question Teller

What's it for?
Children learn about linking two question starter words to develop the range, depth and style of questions that they ask.

How does it work?
Provide each child with a copy of the 'Question Teller'. Cut the outline and fold to create a traditional pyramid-style teller. Ask children to work in pairs to use their 'Question Teller' for question-producing about things of common interest around the science topic or theme being studied.

What to expect when it is going well?
Children produce questions that are further enriched and extended beyond the initial first question.

Tools created for the Great Science Share for Schools' Question Makers.

How to embed the QuBuild Process in classroom practice

Children's Question Frame

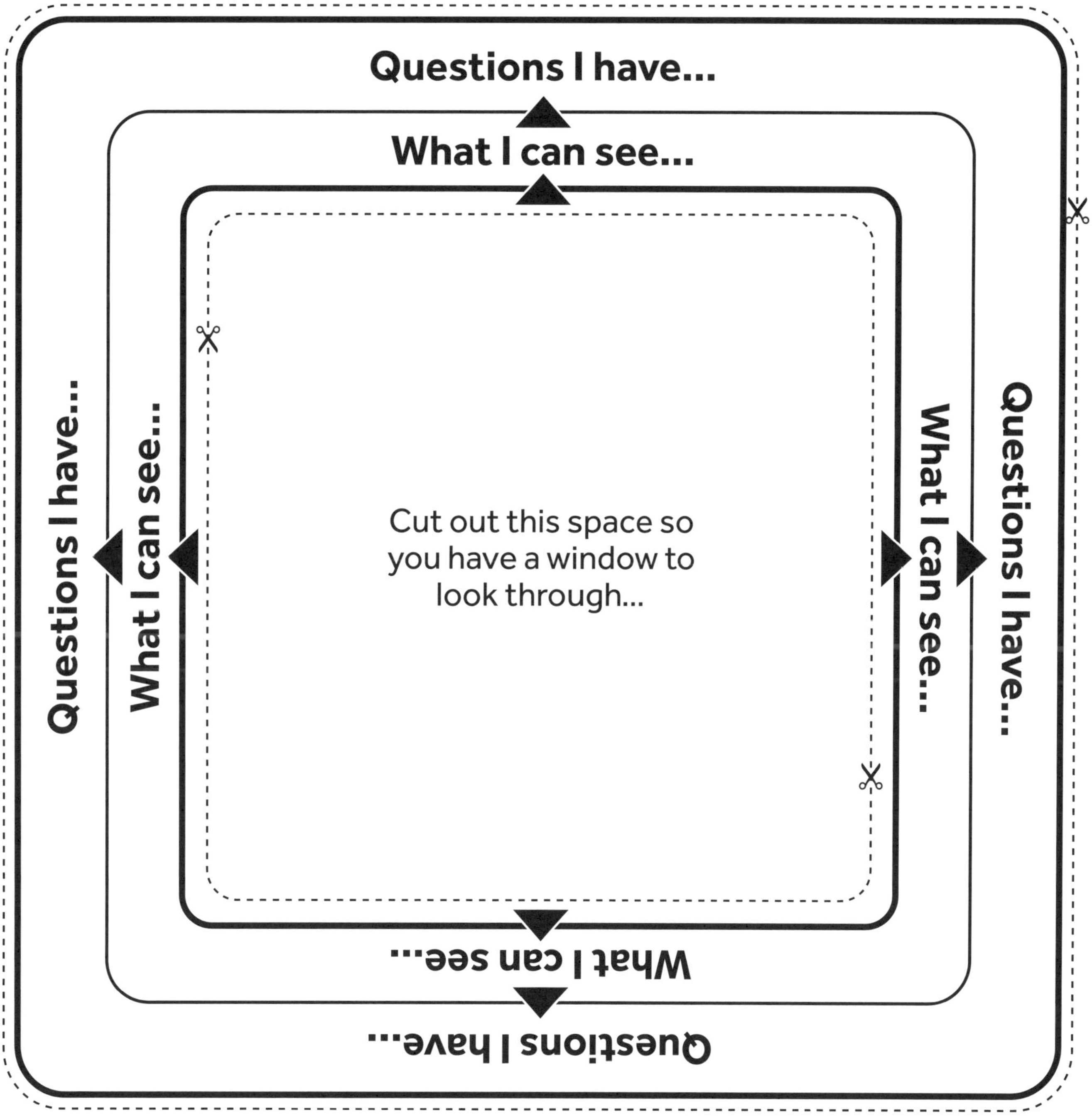

```
Questions I have...
What I can see...

Questions I have...    What I can see...    What I can see...    Questions I have...

Cut out this space so
you have a window to
look through...

What I can see...
Questions I have...
```

Step 1: Question-producing

Children's Wonder Bubbles

I wonder why...?	I wonder what happens when...?	I wonder how does...?
I wonder how would...?	I wonder can you...?	I wonder what if...?
I wonder how come...?	I wonder which is...?	I wonder why does...?
I wonder whether...?	I wonder what...?	I wonder if there is...?

Children's Question Teller

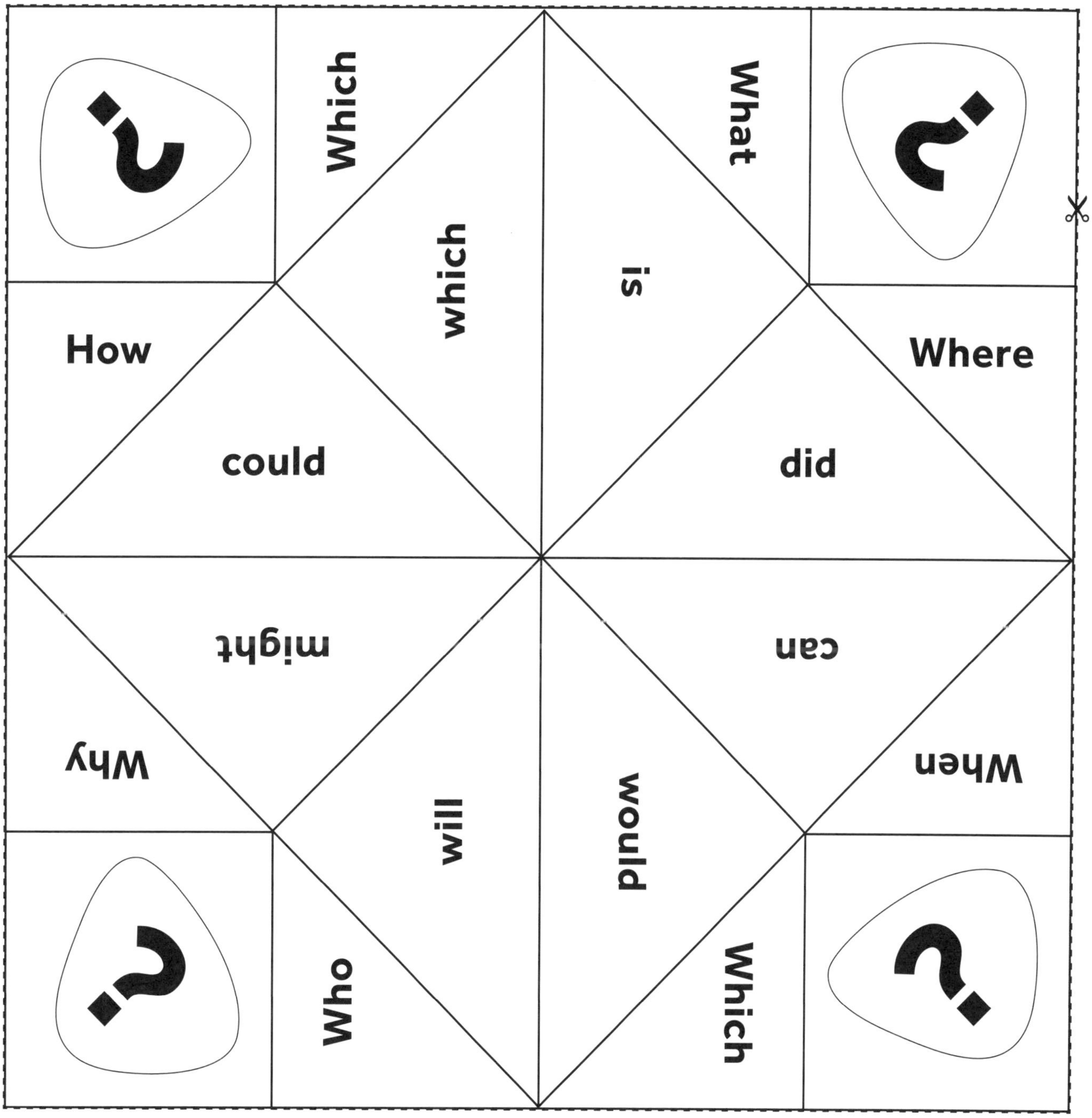

?	Which · which	What · is	?
How · could			Where · did
Why · might			When · can
?	Who · will	Which · would	?

Step 1: Question-producing

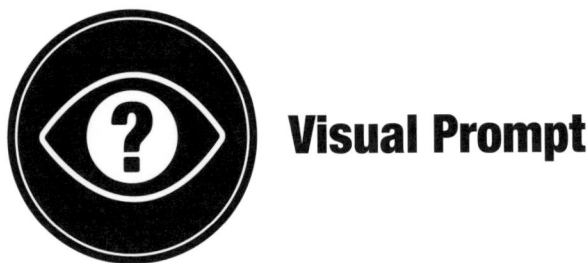

Visual Prompt

Question-producing icon

What's it for?
The Question-producing icon is an image that can be used to scaffold and support children to independently engage in step one of the QuBuild Process.

How does it work?
The Question-producing icon should be regularly used and associated with the first step of the QuBuild Process. Each step of the QuBuild Process has an image that acts as a visual prompt for the step. The teacher should make the icon visible in the classroom, and make reference to it when reminding the children of how to question-produce. Teachers may consider displaying the icon on the white board, wall displays or as a table top hand-out.

What to expect when it is going well?
Children should be familiar with the need to produce further questions and not just stop at the first one. Children will become better at question-producing and increasingly work independently of teacher support.

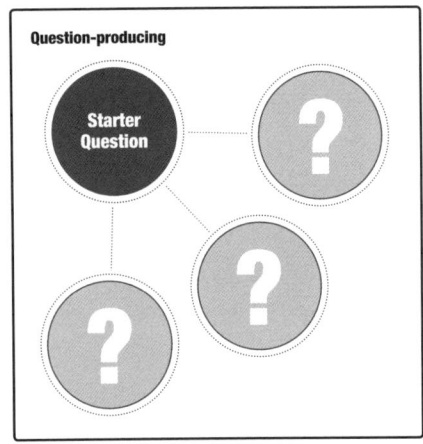

Every child should have the chance to have their question included and listened to.

Question-producing

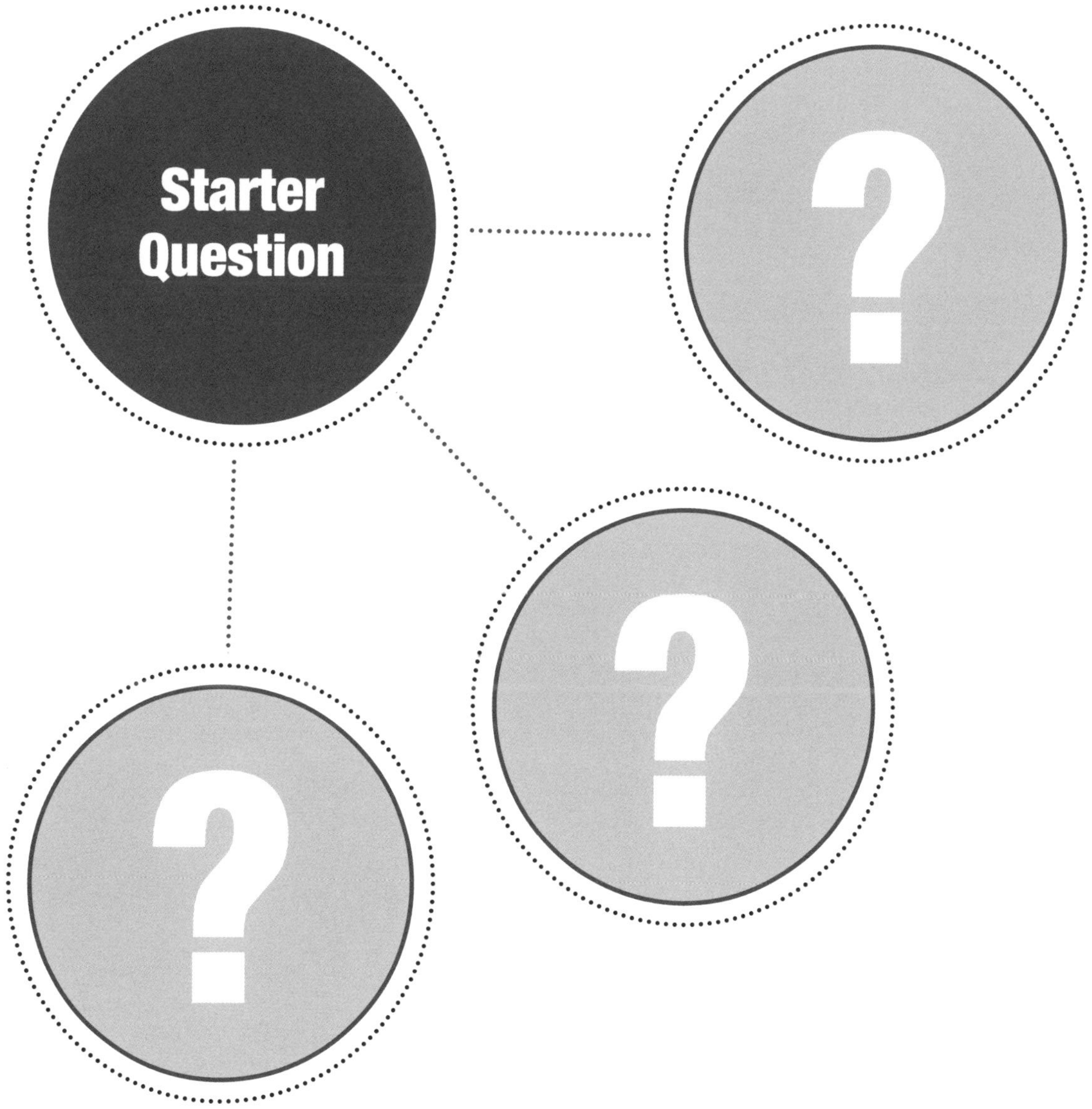

- **Starter Question**
- ?
- ?
- ?

Step 2
Question-handling: children working with questions

Why do we need this step?

Children learn that scientists do not generate questions in isolation, but rather it is more usual for scientists to work in teams than on their own. These teams of scientists are often located in different places, forming a scientific global community.

What does this step involve?

By talking about and describing many and varied questions children develop an understanding that different questions are fit for different purposes. They start to analyse questions to consider purpose, relevance and appropriateness.

What is question-handling about?

This step results in children knowing that all questions are valuable and are useful for different purposes. Children use this step to identify the similarities and differences in their questions. They have the chance to sort questions into groups they choose as well as groups their teachers may suggest. Children engage in reading and debating their questions with their peers, sort their questions into groups and explain the grouping choices that they have made.

As a result, children move away from thinking that, "My question is the most important question because I asked it!". Question-handling emphasises that all questions should be worked with, thereby encouraging inclusion of questions that otherwise might get left behind.

By using this step, the expected impact will be:

From	To
Children's questions being left unconsidered	Children's questions leading to the co-creation of shared questions
Children being told which question they will investigate	Children working with their own questions
Children thinking that their own question is the best one	Children having co-ownership of the question to be explored
Children having little autonomy in which questions they investigate	Children self-auditing their own questions, beginning to discriminate for themselves those worthy of taking forward towards investigation

What knowledge do children need to improve their question-handling?

> Children will benefit from knowing:
>
> - the difference between an open and closed question
> - the difference between a question and a statement
> - the difference between independent variables and dependent variables
> - the differences between the five enquiry types
> - that some questions can be answered by science and some questions cannot be answered by science

Classroom Resources provided to support Question-handling

Introductory Activities
- Question Shuffle
- Question Catalogue
- Variable Spotter

Learning Tools
- Question Sort
- Variable Highlight

Visual Prompt
- Question-handling icon

Step 2: Question-handling

Introductory Activities

Question Shuffle

What's it for?
Children identify features of questions. Children can compare and contrast different questions in relation to the question type and quality and not simply the content or context.

How does it work?
Ahead of the activity, prepare sets of the exemplar 'Question Cards' provided. Ask the children to work in small groups to read, handle and shuffle the cards.

They should come up with a rule that they will use to separate the questions into two groups and should note the classification rule they have used.

Ask the children to reshuffle and regroup the questions repeatedly into different groups. Ask the children to come up with different, new rules to sorting the questions into two groups. They should note each new classification rule. Ask the children to see how many times they can reshuffle and regroup the question cards. Encourage them to be creative, and accept all grouping ideas. Compare and share different ideas for grouping from different groups.

What to expect when it is going well?
Children can group and classify questions using their own ideas. They appreciate there are multiple ways to sort questions for different purposes and reasons.

Teacher Notes

There are no correct answers to this open-ended sort activity whereby free choice is encouraged and enjoyed.

Examples of classification rules might include:

- I know the answer/ I don't know the answer.
- I would like to know the answer/ I am not interested in the answer.
- I can ask someone the answer/ I can find out the answer.
- …etc.

Children's Question Cards

Where do woodlice live?	Is it better for fitness to run steady long distances or fast short distances?	What is the most common colour of trainers in the school?
Where do snakes live?	What time can children in our class go to lunch?	Which woodlouse from the school grounds is the biggest?
What happens to pulse rate when you exercise?	Do all animals sleep every day?	Where can I find out more information about spiders?
Are snakes horrible?	Are there more woodlice than snakes in the local woods?	When is the best time of the day to exercise?
How many legs does a spider have?	How do woodlice respond to changing temperature?	Where is the sunniest spot on the playground?
How many children in the school have a pet snake?	What is the best material to wear for exercise?	Why do animals need to eat food?
What kind of animal is a woodlouse?	Is it cruel to keep animals in a zoo?	What do snakes eat?

Step 2: Question-handling

Question Catalogue

What's it for?
Children learn to appreciate that there are many types of questions. The 'Question Catalogue' introduces new vocabulary by sharing definitions to six question types which will make Question-handling quicker on going.

How does it work?
This is a matching, sorting and classifying activity. Teachers can organise the activity dependent on their classroom set up, some teachers may choose to give children the list of six question types and display one question at a time on the board, whilst others may prefer cards for children to handle. The basic principle is for groups of children to discuss, debate and compare the choices made.

What to expect when it is going well?
Children are familiar and confident to use the language in the descriptors of the six question types provided in the resources that follow.

Teacher Guidance (Correct Responses)

Question	Question type
Is copper magnetic?	Closed question
Are spiders friendly?	Opinion question
What is climate change?	Fact-find question
How fast does ice melt?	Empirical question
How much does your heart rate change when you do different exercises?	Cause-effect question
How certain are we that plastics harm the planet?	Trust question
Is gravity a force?	Closed question
Is chocolate tasty?	Opinion question
Who first used the term gravity?	Fact-find question
What happens when light shines on objects?	Empirical question
How long does it take for water to filter through sandy soil?	Empirical questions
How much does the length of an elastic band change the pitch of the sound when plucked?	Cause-effect question
How can we be sure that medicines work?	Trust questions
Can we ever make oil and water mix?	Empirical questions
How can different pitches of sound be produced?	Empirical questions

Children's Question Type Cards

- Closed questions
- Empirical questions
- Opinion questions
- Cause-effect questions
- Fact-finding questions
- Trust questions

Step 2: Question-handling

Children's Question Cards

Is copper magnetic?	How certain are we that plastics harm the planet?
Are spiders friendly?	How long does it take for water to filter through sandy soil?
What is climate change?	Is gravity a force?
Can we ever make oil and water mix?	Is chocolate tasty?
How fast does ice melt?	Who discovered gravity?
How much does your heart rate change when you do different exercises?	What happens when light shines on objects?
What happens when milk mixes with washing up liquid?	How much does the length of an elastic band change the pitch of the sound when plucked?
How can we be sure that medicines work?	How can different pitches of sound be produced?

Variable Spotter

What's it for?
Children identify the variables within cause-effect questions.

How does it work?
Children are already familiar with the cause-effect question type from the activity 'Question Catalogue'. Variable Spotter focuses on identifying the two variables within this question type. Teachers will need to introduce to children the specific scientific terms of independent and dependent variables. As an introductory activity children are given prepared questions. In pairs, children look for or spot the independent variable (what is changed in the enquiry) and the dependent variable (what is measured or observed in the enquiry). Some of the questions have both variables present, some only one variable and some might not be a cause-effect question type at all. Initially the presence of the variable is recorded as a yes or no in the recording table.

As an extension the teacher might ask children to describe each variable and encourage discussion of whether the variable is clear enough or could be improved.

What to expect when it is going well?
Children can determine the independent variable from the dependent variable. Children can recognise the number of variables in a question.

Step 2: Question-handling

Children's Variable Spotter Questions

Questions	Can you spot the independent variable? Yes/No What is it?	Can you spot the dependent variable? Yes/No What is it?
Does the size of the parachute affect the time it takes to fall?		
Will changing the temperature of the water affect how long it takes the sugar to dissolve?		
How does the height of the plant change?		
Which is the best spinner?		
What happens to pulse rate when I exercise?		
Is a thicker paper towel the best?		
Do bigger seeds grow into taller plants?		
Are expensive washing-up liquids always better than cheaper ones?		

Teacher Guidance (Correct Responses)

Questions	Can you spot the independent variable? Yes/No What is it?	Can you spot the dependent variable? Yes/No What is it?
Does the size of the parachute affect the time it takes to fall?	Yes Size – surface area would be more specific	Yes Time
Will changing the temperature of the water affect how long it takes the sugar to dissolve?	Yes Temperature	Yes Time
How does the height of the plant change?	No	Yes Height
Which is the best spinner?	No	No
What happens to pulse rate when I exercise?	Yes (Exercise or not)	Yes (Pulse rate)
Is a thicker paper towel the best?	Yes (Thickness)	No
Do bigger seeds grow into taller plants?	Yes (Seed size or length)	Yes (Plant height)
Are expensive washing-up liquids always better than cheaper ones?	Yes (Cost)	No

Step 2: Question-handling

Learning Tools

Question Sort

What's it for?
Through the introductory activities of 'Question Shuffle' and 'Question Catalogue' children learn to handle and describe a given set of sample questions. The 'Question Sort' tool is designed to make the question-handling step more systematic by providing children with prompts to structure their peer-to-peer discussions.

How does it work?
The 'Question Sort Flash Cards' provide closed questions that children can ask of any question they are handling. The question cards are designed to be reusable and a resource the children have access to easily and when required. When children use each 'Question Sort Flash Card' it will result in the number of questions they handle becoming fewer and more focused on an enquiry that can follow.

What to expect when it is going well?
Children can sort and handle their own questions systematically with increasing independence from the teacher.

Question Sort Flash Cards

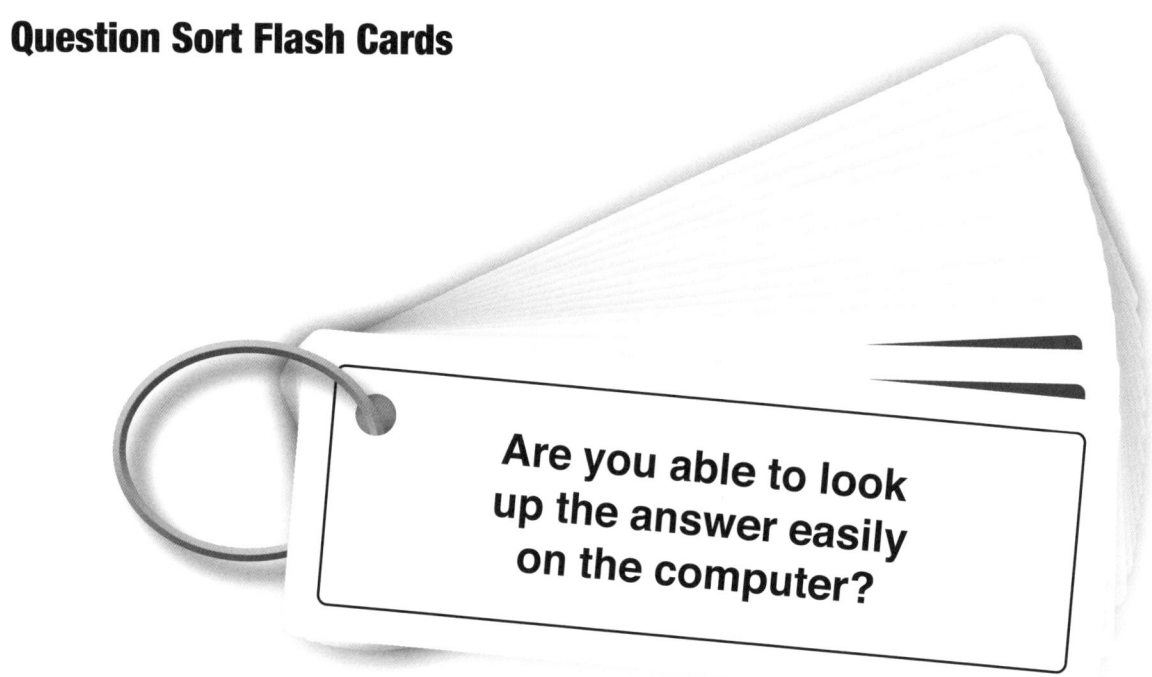

Children's Question Sort Flash Cards

Cut out each box to create a question classification flash card. Punch a hole to the end of each card. Place the cards to a flash ring to make a reusable set of prompts when question-handling.

- ○ Are you able to look up the answer easily on the computer?
- ○ Do you need to make something new to answer the question?
- ○ Can evidence be gathered by experiment in the classroom?
- ○ Will you be able to answer the question by doing a survey?
- ○ Is it a question that a scientist or others might have more evidence about?
- ○ Do you already know the answer to the question?
- ○ Is it a scientific question?
- ○ Do you need to ask a specialist about this question?
- ○ Does the question have two variables?
- ○ Will your question give results for a graph?
- ○ Is the question open?
- ○ Do you need to use scientific measuring equipment to answer your question?
- ○ Does the question make you curious?
- ○ Will you be able to answer the question by reading information text?

Variable Highlighter

What's it for?
Children use the 'Variable Highlighter' learning tool to identify the independent and dependent variables when handling any Cause-effect question.

How does it work?
The 'Variable Spot' activity guided children through question examples to determine independent and dependent variables. This learning tool provides a scaffold that children can use with minimal teacher support that can be applied to any cause-effect question

Children choose two highlighter pens of different colours for this learning strategy. Once the highlighter colours have been selected use these colours consistently across the school and year on year. One colour pen represents the independent variable (what is changed in the enquiry) and the other colour pen represents the dependent variable (what is measured or observed in the enquiry). Children are asked to highlight the variables using the representative colour.

What to expect when it is going well?
Children sort and handle any question by understanding the variables are present. Children are able to identify single variable questions as different to questions with two variables, and different to those that have no variables and are not cause-effect questions.

Equipment needed
Two different coloured highlighter pens per group.

QuBuild Science Keywords

Vocabulary Keywords	Definition
Statement	A statement is the most common type of sentence. There are three other sentence types: questions, exclamations and commands.
VARIABLES	
Independent variable	A variable that can be changed. In a fair test, only one independent variable is changed.
Dependent variable	A variable that is measured in a scientific investigation and which changes as a result of the independent variable changing.
ENQUIRY TYPES	
Pattern Seeking	Making observations and measurements to explore natural events where there are variables that they can't easily control. Seeking to identify patterns in the measurements, which may lead to other investigations in an effort to try to explain why a particular pattern occurred.
Observation over time	Observing and measuring events and changes in living things, materials and physical processes or events over time. These observations may take place over time spans of minutes or hours, up to several weeks or months.
Fair and Comparative Testing	Identifying the effect of changing one independent variable on another dependent variable whilst attempting to keep other variables the same. Gathering data that might inform predictions and further tests. In comparative tests this involves comparing one event with another and identifying different outcomes. With fair tests children look to identify a causal relationship between two variables.
Research by Secondary Sources	Using a range of secondary sources (books, websites, articles, people, videos etc.) to gather evidence to answer questions. Looking for patterns in the information they collect, evaluating the reliability and trustworthiness of the evidence they collect when drawing conclusions.
Identifying and Classifying	Identification is the process of using observable differences to name something and classification is organising things into groups based on observations, features and characteristics.

Step 2: Question-handling

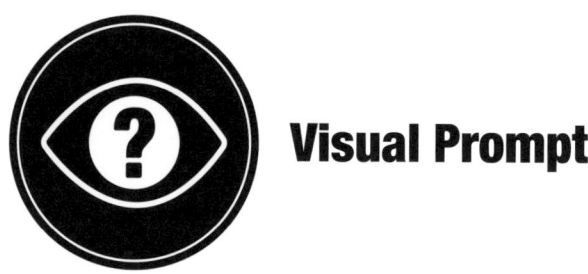

Visual Prompt

Question-handling icon

What's it for?
The Question-handling icon is an image that can be used to scaffold and support children to independently use the second step of the QuBuild Process.

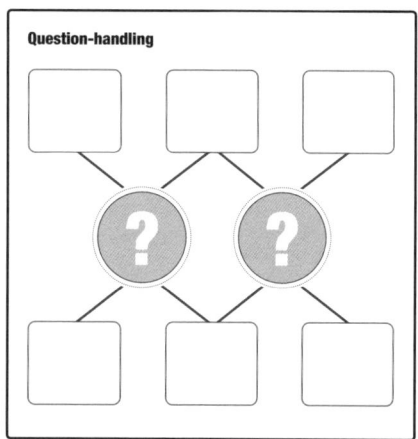

How does it work?
The Question-handling icon is to be regularly used and associated with the second step of the QuBuild Process. Each step of the QuBuild Process has an image that acts as a visual prompt to support children to recall the relevant thinking. The teacher should make the icon visible in the classroom, and make regular reference to when reminding the children of how to question-handle. Teachers may consider displaying the icon on the white board, wall displays or as a table top hand-out.

What to expect when it is going well?
Children should be familiar with different types of questions and be able to group questions by identifiable features. Children can explain the similarities and differences when handling questions.

Teacher Notes

The icon should be used by identifying two things about the questions that are the same, and two things that are unique to each question.

How to embed the QuBuild Process in classroom practice

Question-handling

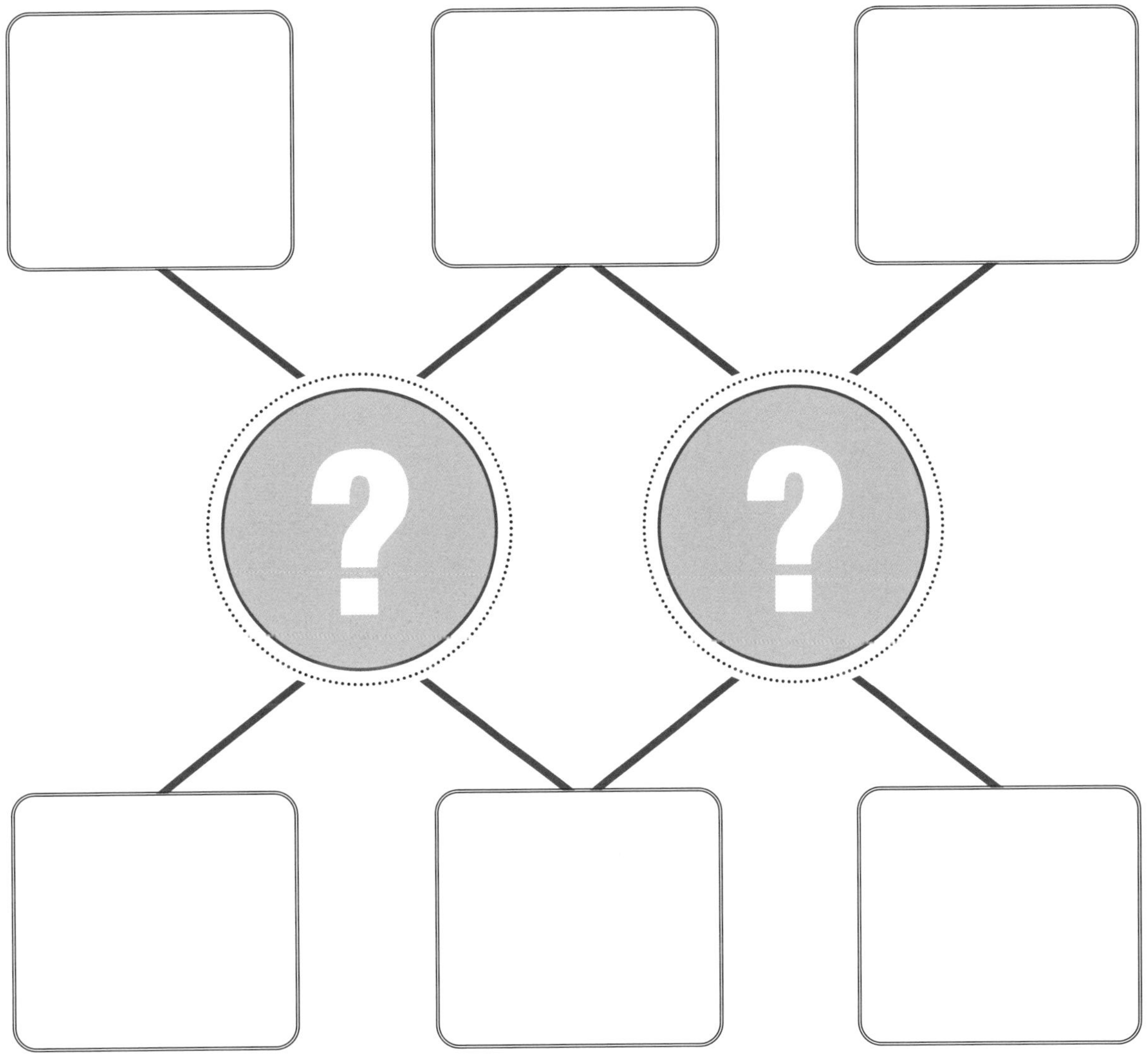

Step 3
Question-improving and choosing: children refining and building scientific questions

Why do we need this step?

This step supports children to know that scientists invite others to comment on and review their work. This is called 'peer-review' and is a key part of the way scientists work scientifically. It is a world-wide activity that scientists regularly use to improve the rigour, trustworthiness, and novelty of their research. As a result of this step children will understand the significance between the quality of the scientific question they ask, and the quality of the evidence they generate when investigating it. Step 3 of the QuBuild Process reinforces how scientists develop explanations and conclusions based on evidence.

What does this step involve?

Children develop questions into better scientific questions to improve the quality of the evidence that they gather. Children focus on how they can improve and choose their questions before going on to doing the investigation.

Improving involves reflecting and working on the construction and phrasing of the question as this will shape the kind of evidence the children will gather.

Choosing involves reviewing whether the kind of evidence the children gather is fit for purpose. By learning to prioritise the best questions to investigate they will develop understanding and skill in identifying those where conclusions can be drawn that move their understanding and knowledge on. When doing this, it also gives children the opportunity to think about how their situational context - the space, time, and resources they have affects the scientific question they investigate.

What is question-improving about?

This step results in children knowing that all questions should be well-thought out before they start their enquiry. Having produced and handled their questions children can now collaboratively agree on those to be developed into an enquiry, discuss and justify their selection, and make 'tweaks' to improve the quality of evidence anticipated, recognising validity, reliability and trustworthiness as essential.

By using this step, the expected impact will be:

From		To
Children lack autonomy and rely on the teacher to provide a 'good' question to investigate		Children having the knowledge to improve their own questions and embrace the independence to do so
Children having little or no time to review and justify the questions they explore		Children reviewing their own and others' questions in an analytical way
Children only considering the quality of the evidence in the reviewing stage of their enquiry		Children recognising the need to plan for high-quality evidence
Children feeling inadequate because 'their' question has been put to the side		Children appreciating that by working together a 'better' question for the purpose has been taken forward for the enquiry

What knowledge do children need to get better at question-improving and choosing?

> Children benefit from knowing:
>
> - scientific knowledge is tentative and subject to change based on new evidence or new interpretation of existing evidence
> - that repeat readings and controlled variables improve the reliability of the enquiry
> - that considering sample size and range improve the validity of the enquiry
> - scientists look to find cause and effect relationships

Classroom resources provided to support Question-improving and choosing

Introductory Activities
- Question Pick
- Question Switch
- Question Tweak

Learning Tools
- Question Checker

Visual Prompt
- Question-improving and choosing visual icon

Step 3: Question-improving and choosing

Introductory Activities

Question Pick

What's it for?
Children learn to appreciate that whilst there are many types of questions each have advantages and disadvantages. There is often not a perfect fit but rather a compromise in making the best choice for the intended enquiry approach.

How does it work?
In the 'Question Catalogue' activity children have learnt the names and features of different question types. 'Question Pick' is a way for children to revisit the six question types and to begin to choose questions based on thinking about and discussing the advantages and disadvantages of each type. The 'Question Pick' table is provided to structure this activity. The table can be provided for small groups to complete or led as a whole class discussion by the teacher. Suggested responses are included but are not exhaustive.

What to expect when it is going well?
Children become fluent in articulating the preferred question type for a particular topic or purpose.

QuBuild Question Keywords

Question type	Description of the question type
Closed questions	Require a 'Yes-No' answer. Open questions are all other questions that do not result in a 'Yes-No' answer
Opinion questions	Cannot be answered by scientific enquiry, they are about what people think, feel or believe
Fact-find questions	Can be answered using an information search, agreed definition, generally descriptive
Empirical questions	Are based on observing a change or effect, not a theory
Cause-effect questions	Evaluate the effect of one thing on another. Includes at least two variables
Trust questions	Evaluate the certainty of information and its trustworthiness

Children's Question Pick

Question Type:	
Advantages of this type of question	Disadvantages of this type of question

Question Pick (Possible Responses)

Question Type: Closed Questions	
Advantages of this type of question	**Disadvantages of this type of question**
Are easy to respond to	Are too simplistic or superficial
Are easy to compare answers, e.g. from Yes-No questions or surveys	Lack depth
Give straight-forward, simple, or less confusing responses. The data collected is easy to analyse so that trends in the evidence can be reviewed	Give little indication of whether or how to trust the answer that is given. A yes or no could be a guess

Question Type: Opinion Questions	
Advantages of this type of question	**Disadvantages of this type of question**
Give insight to what people think	Cannot be tested
Are inclusive and value all ideas	Have limited relationship to evidence or data
Gather people's reactions to an observation or wondering	Create data that is more difficult to analyse statistically or with numbers

Question Type: Fact Finding Questions	
Advantages of this type of question	**Disadvantages of this type of question**
Are quick to answer, e.g. internet search	Can lead to the assumption that the first or most frequently given answer is the right answer
Value the knowledge that other people have provided	Don't recognise the source of the information
Are used to check understanding	Can lead to the assumption that what's found on the internet is accurate, unbiased and trustworthy

Question Type: Empirical Questions	
Advantages of this type of question	**Disadvantages of this type of question**
Are less biassed as the observations provide the direct evidence	Require time for enough evidence to be gathered over time, which can be challenging in classroom spaces
Are reliable as it is real experience and not an opinion or idea	Can lead to evidence being gathered where human error is more likely
Support the question-asker to use and apply systematic methods to gather data over time	Require equipment which may be limited in schools

Question Type: Cause-Effect	
Advantages of this type of question	**Disadvantages of this type of question**
Are testable	Are thought to be the main or only type of scientific question to provide good evidence
Make it clear in the question what data is to be collected	Are oversimplified
Are easier to draw a conclusion from	Lead to incorrect conclusions being drawn if the causation (the cause) and correlation of evidence is easily confused

Question Type: Trust Questions	
Advantages of this type of question	**Disadvantages of this type of question**
Are helpful to recognise and appreciate uncertainty	Are open-ended with no clear answer
Are helpful to identify where bias may be influencing observations or evidence	Have multiple evidence sources and need rigorous analysis to draw conclusions
Allow us to see how scientific understanding can change over time	Are difficult to communicate and give little indication of how ideas have changed over time

Step 3: Question-improving and choosing

Question Switch

What's it for?
Children learn how to adapt questions by switching the type of question whilst keeping what the question is about the same. Children start to think about the quality of the question by considering if changing its type could improve the evidence they might gather. With teacher support children begin to appreciate that adapting questions can lead to different and better scientific outcomes. They will move from investigating the first question they have come up with, to considered questions fit for enquiry.

How does it work?
Previously in 'Question Shuffle' children became familiar with features of questions and in 'Question Catalogue' they became familiar with six question types. 'Question Switch' is an activity to build upon this prior understanding.

The 'Question Switch' template supports children to question-improve by switching a range of sample questions from one type to another. It is important that they are reminded to keep what the question is about the same.

The 'Children's Question Switch Template' has the two rules visible, and teachers should demonstrate its use by modelling the approach. 'Question Switch - Teacher Guidance' provides examples of where a question has been switched from one type to another so that it is easy for illustrating the approach. It is important that children appreciate there is no right answer in this activity and that its value comes from the increased confidence and ability to work with questions to explore alternative outcomes.

What to expect when it is going well?
Children will improve their scientific questions by describing and explaining how they have switched it and how this has changed the evidence they would gather. Children will identify question features, such as the question stem, the question type, and increasingly start to recognise how changing the type impacts on the question quality and the scientific evidence they gather. They would start to appreciate which are the 'better' questions for their investigations.

Is copper magnetic?

Switch it
1. Choose a new question type
   From a closed question to an empirical question
2. Keep what the question is about the same
   Magnetism

Which materials in the pot are magnetic?

Teacher Notes

Provide pairs of children with the set of 'Children's Question Switch Sample Questions' and a 'Children's Question Switch Template'. The teacher should encourage discussion as the children create new and discuss which are better questions.

Children's Question Switch Template

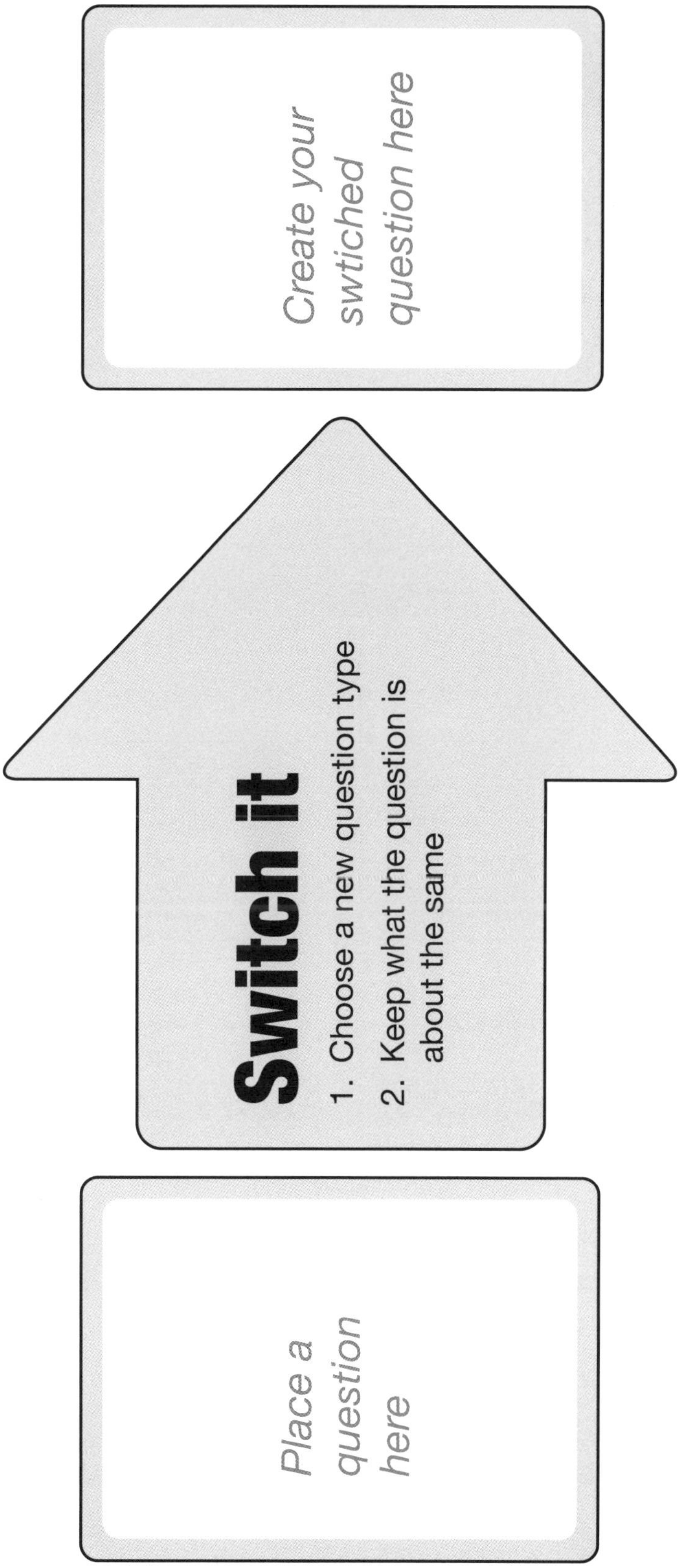

Create your swtiched question here

Switch it

1. Choose a new question type
2. Keep what the question is about the same

Place a question here

Step 3: Question-improving and choosing

Children's Question Switch Sample Questions

| Is gravity a force? **A closed question** | Is a cactus a plant? **A closed question** |

| Is chocolate tasty? **An opinion question** | Are spiders friendly? **An opinion question** |

| How fast does ice melt? **An empirical question** | What happens to an egg dropped into water? **An empirical question** |

| When was climate change first discussed by politicians? **A fact-finding question** | What is the force of gravity on the moon? **A fact-finding question** |

| How does your heart rate change when you do different exercises? **A cause and effect question** | How does the length of a ruler change the pitch of the sound when twanged? **A cause-effect question** |

| How sure are we that plastics harm the planet? **A trust question** | How sure are we that eating less meat is better for us? **A trust question** |

| Your first question **A closed question** | Your first question **A fact-finding question** |

| Your first question **An open question** | Your first question **A cause and effect question** |

Question Switch – Teacher Guidance

Some possible examples are included for teacher guidance to encourage children if needed. You should anticipate and celebrate the creativity of the new questions the children produce when they engage in 'Question Switch'.

Question type	Question	New question	New Question type
Opinion	Are spiders friendly?	Which spiders are safe to handle?	Fact-finding
Closed	Is copper magnetic?	Which materials in the pot are magnetic?	Empirical
Empirical	How fast does ice melt?	How does the shape of the ice affect the time it takes to melt?	Cause-Effect
Fact-finding	When was climate change first discussed by politicians?	Who should we trust to explain climate change?	Trust
Cause-effect	How much does your heart rate change when you do different exercises?	Do hearts look great on Valentine cards?	Opinion

Step 3: Question-improving and choosing

Question Tweak

What's it for?
During the 'Question Tweak' activity children are likely to have produced a range of questions that are all of a similar type, share a similar subject but are of different quality. Children will require support from the teacher to notice the adaptations that improve questions and to be able to use these in the future. This activity should be a collaborative teacher-children discussion.

How does it work?
The resource provides worked examples for class discussions. The teacher should display two questions of the same type and guide the children to identify and discuss the changes that have been made. The 'tweaks' should be described in terms that the children best understand at a level the teacher considers appropriate. By noticing the changes made to the question, the routine of identifying how questions are tweaked becomes a future tactic for children to use more independently when question-improving.

What to expect when it is going well?
Children will develop a range of tactics for question-improving. Children are able to explain why one question is better than another.

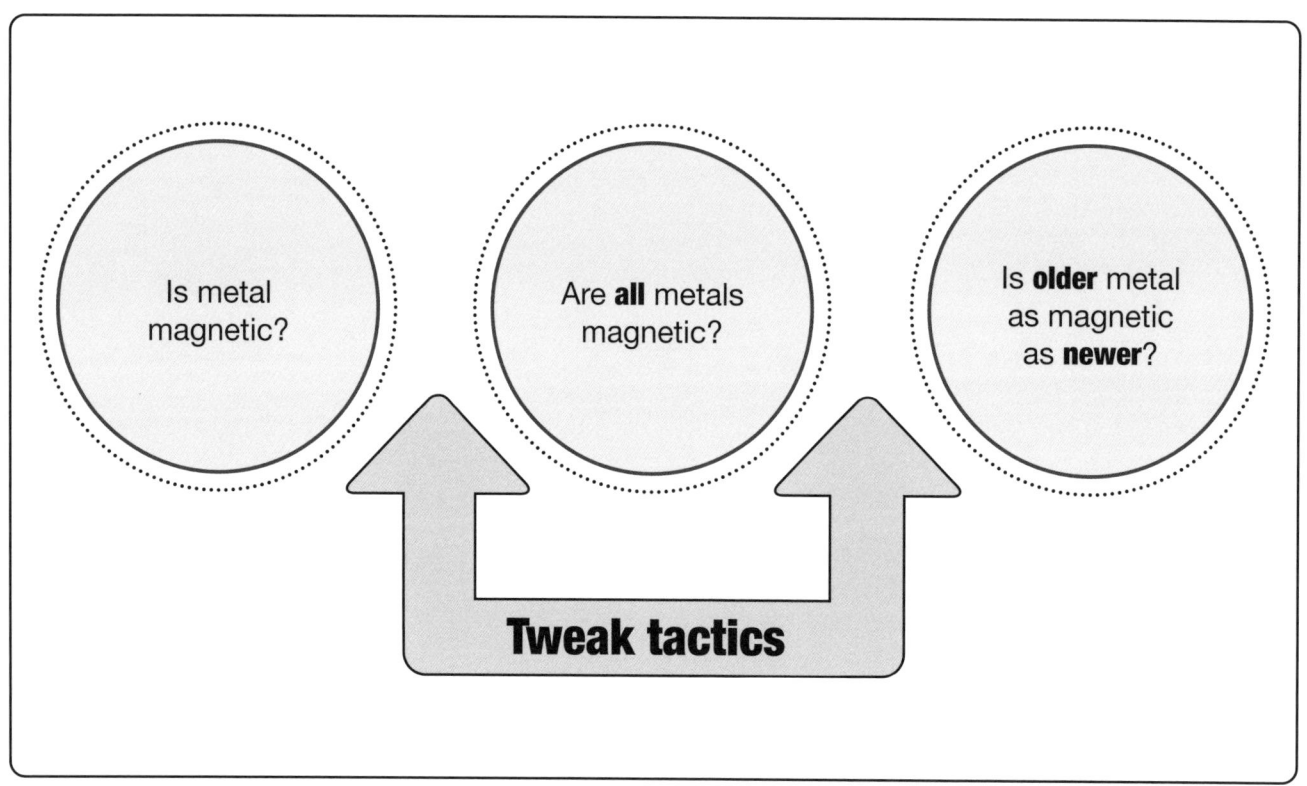

Teacher Notes

There are different ways to improve questions. By demonstrating the process with the children, you will find that some of the following tactics can support the development of the question.

TACTIC 1
Add an adjective, e.g. older, newer, shorter, longer.
E.g. This would tweak the question from 'Is copper magnetic?' to 'Is **older** copper as magnetic as **newer**?

TACTIC 2
Change the question stem.
E.g. This would tweak the question from 'Are spiders friendly?' to '**Can** spiders be a friend?'

TACTIC 3
Insert the source/ person whose opinion it is.
E.g. This would tweak the question from 'Are all spiders friendly?' to 'Do all children **in our class** think spiders are friendly?

TACTIC 4
Insert the situation where the observations are being made.
E.g. This would tweak the question from 'How fast does ice melt?' to 'Does ice melt faster **outside or inside**'

TACTIC 5
Insert the source of the facts found, the author of the facts found and/or the organisation presenting the facts found.
E.g. This would tweak the question from 'What is Climate change?' to 'What did people think about Climate Change **in the 1900's**?', 'What does your **school website** say about climate change?', 'What does David Attenborough say climate change is?', 'Are the facts you are finding about Climate change trustworthy?'

TACTIC 6
Find the independent variable, dependent variable or control variable.
E.g. This would tweak the question from 'How much does your heart rate change when you do different exercises?' to 'How many **beats per minute** does your heart rate make during **jumping on the spot exercise**s?', 'How does the longer you run on the spot affect *your* heart rate?

TACTIC 7
Insert a number word, e.g. 'all', 'some', 'a few'.
E.g. This would tweak the question from 'How certain are we that plastics are bad?' to 'How certain are we that **all** plastics harm the planet?' 'How certain are we that all plastics are bad for **all** the wildlife on our planet?'

Step 3: Question-improving and choosing

Children's Question Tweak

Observing the world	Improved questions for scientific evidence gathering	Spot the tweak... What has been added to improve the question?
Is copper magnetic?	Is all copper magnetic? Is older copper as magnetic as newer copper?	
Are spiders friendly?	Do all children in our class think spiders are friendly?	
How fast does ice melt?	Does ice melt faster outside or inside?	
What is climate change?	What did people think about climate change in the 1900's? What does your school website say about climate change? What does David Attenborough say climate change is? Are the facts you are finding about climate change trustworthy?	
How much does your heart rate change when you do different exercises?	How many beats per minute does your heart rate make during jumping on the spot exercises? How does the longer you run on the spot affect your heart rate	
How certain are we that plastics are bad?	How certain are we that all plastics harm the planet? How certain are we that all plastics harm the wildlife on our planet?	

Learning Tools

Question Checker

What's it for?
Through the introductory activities of 'Question Pick' and 'Question Tweak' children have begun to appreciate that questions can be improved. They are increasingly used to giving and receiving constructive feedback to their peers. The 'Question Checker' is a tool to check how satisfied the children are with their question.

How does it work?
It enables the children to check whether their question is fit for purpose. The 'Question Checker' tool is a simple checklist of five prompts. Each prompt provides a challenge and suggestion for improvement to the question.

Children do not need to use all five checks all of the time. Encourage children to say which check they have used and to share annotated new improved questions.

What to expect when it is going well?
Children will be able to monitor and audit questions, including their own. By using the checklist children can describe how their question meets a quality check. Children will celebrate when the quality check stage supports an even better question as the QuBuild Process encourages critical reflection and analysis of questions generated.

Step 3: Question-improving and choosing

Children's Question Quality Checklist

Question Quality Checklist

☐ **1.** Remember **scientific evidence can be unreliable** by making claims in relationships that are not true. Check that your question includes the right cause, effect and control variables.

☐ **2.** Remember **scientific evidence can be uncertain**. Check that your question includes what will be measured, how the readings will be measured.

☐ **3.** Remember **scientific evidence might not be trustworthy**. Check that your question includes the sample size and repeat readings involved.

☐ **4.** Remember **scientific knowledge is sometimes updated with new evidence**. Check that your question describes when the evidence is to be (or was) gathered.

☐ **5.** Remember **scientific evidence might be interpreted differently by different people**. Check that your question includes who is drawing conclusions from the evidence, e.g. the group or names of the scientist or organisation(s) involved.

How to embed the QuBuild Process in classroom practice

Question Quality Checklist Teacher Guidance

Examples are given below for each of the five checklist points. This enables the teacher to demonstrate with explanation why the quality check is not met and what the improved question might look like.

✓ Checklist	Example question not meeting the Quality Check	A better question improved to meet the Quality Check
☐ **1.** Check that your question includes the right cause, effect and control variables.	**Does the number of shark attacks increase when more ice-cream is sold?** This does not pass the quality check because the cause and effect are not causal. The variables are incorrectly selected.	How does the type of shark affect the number of attacks reported on humans?
☐ **2.** Check that your question includes what will be measured, how the readings will be measured.	**Does an electric car go further in summer than in winter?** This does not pass the quality check because it lacks detail of the evidence to be measured.	How many more laps of a 1km test track will the same electric car go in June 2023 compared to December 2023 when fully charged and captured on a lap counter camera?
☐ **3.** Check that your question includes the sample size and repeat readings involved.	**What is the average age that children start to tie their own shoelaces?** This does not pass the quality check because it does not indicate who is being asked or how many different people are being asked.	What is the average age of shoe lace tying when asking parents of each year 6 group (500 children), every new school year for five years?
☐ **4.** Check that your question describes when the evidence is to be or was gathered.	**How many planets are in the solar system?** This does not pass the quality check because it does not indicate when the information was gathered. In fact the answer would be different fifty years ago to now.	How many planets do space scientists in 2022 agree are in the solar system?
☐ **5.** Check that your question includes who is drawing conclusions from the evidence, e.g. the group or names of the scientist or organisation(s) involved.	**Is eating dark chocolate good for your health?** This question does not pass the quality check as it could be an opinion, or the source providing the answer might represent the findings of a chocolate manufacturer.	What does the 2022 report by food scientists from Sheffield University conclude about the health benefits of eating dark chocolate?

Step 3: Question-improving and choosing

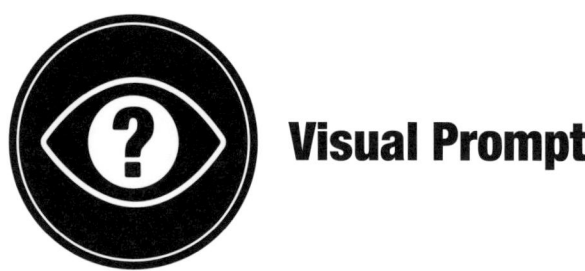

Visual Prompt

Question-improving and choosing icon

What's it for?
The Question-improving and choosing icons are images that can be used to scaffold and support children to independently use the third step of the QuBuild Process.

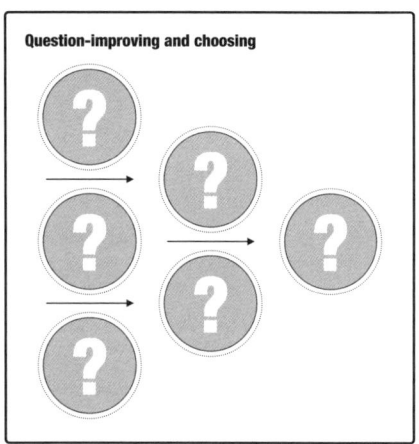

How does it work?
The Question-improving and choosing icon should be regularly used and associated with the last step of the QuBuild Process. Each step of the QuBuild Process has an image that acts as a visual prompting for that step. The teacher should make the icon visible in the classroom, and use it as a reference when reminding the children to question-improve and question-choose. Teachers may consider displaying the icon on the white board, wall displays or as a table top hand-out.

What to expect when it is going well?
Children should be aware that some questions will generate better scientific evidence than others and can justify which questions are of good enough quality to consider as fit for the purpose of the enquiry intended.

Question-improving and choosing

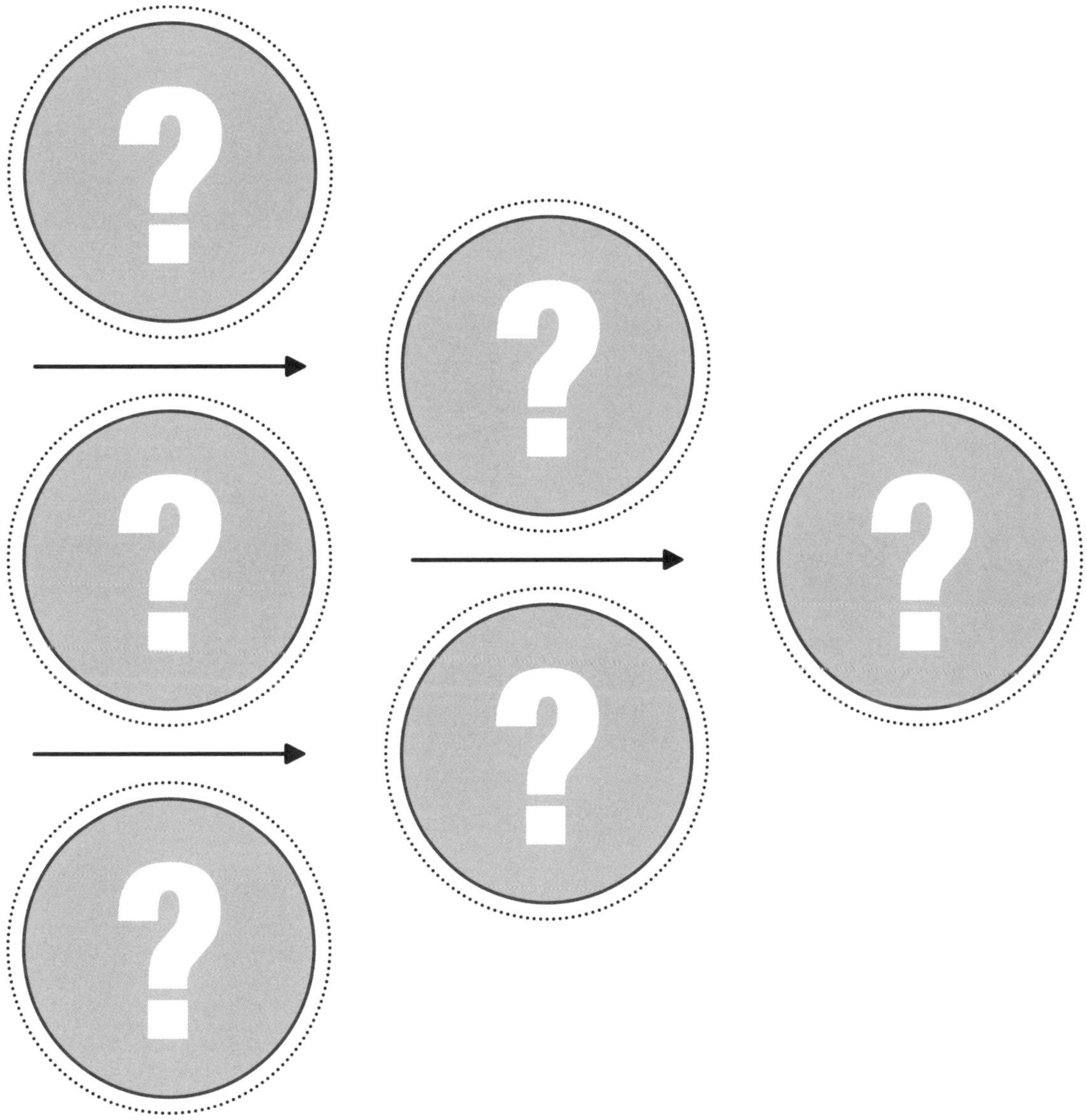

QuBuild Visual Prompts – Step by Step

Step 1

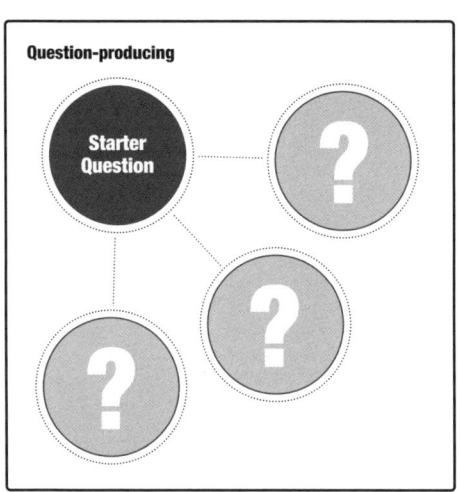

Step 2

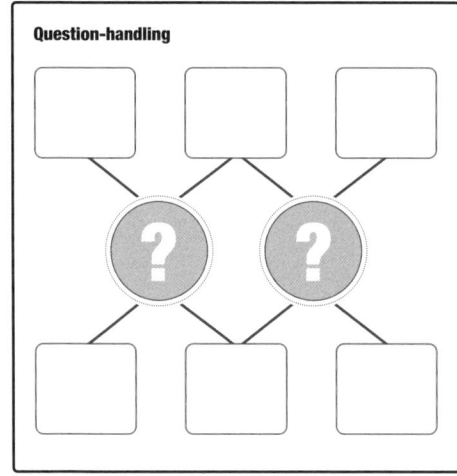

Step 3

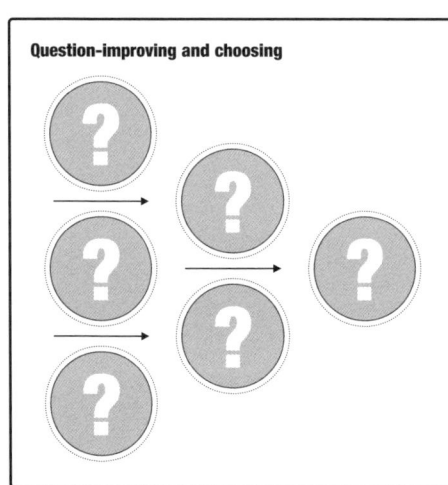

Visual prompts are ways to take thoughts and ideas in our minds and put it into words and drawings so that other people can respond to them.

A future challenge for the teacher will be to facilitate the children to develop their own questions rather than direct them and this like other scientific skills requires practice.

Harlen 2000; Turner 2012

How does the QuBuild Process affect professional learning?

This book has provided stimulus and resources for teachers to develop their professional practice and understanding of how science uses evidence to develop explanations. It is a must-have for all teachers as it addresses the challenge of how to plan for and teach the associated disciplinary knowledge effectively.

What better way to close QuBuild: A guided approach to asking better scientific questions in primary schools than to consider the essential new learning that teachers should take away to achieve that goal.

We considered:

The Fundamental Key Concept

Fundamental to getting better at asking scientific questions is the relationship between quality of evidence and the quality of the question. By improving the question, children can be guided to collecting evidence that will better build their scientific knowledge and deepen understanding of concepts.

The Essential QuBuild Process

There are three steps to the QuBuild Process, Question-producing to elicit more questions; Question-handling to understand and sort questions; and Question-improving and choosing to develop the question ready for enquiry.

The Core Language

When getting better at scientific question-asking, children benefit from explicit teaching of specialised keywords to improve their access to learning. Important to QuBuild is familiarity and common use of terms that underpin the knowledge of effective scientific question-asking.

The Embedded Practice

Regular practice is vital if children are to be given the best chance to use the skill of scientific question-asking with independence and confidence. To achieve independence new thinking requires teacher instruction and modelling followed by scaffolded tools for practice and then finally pupil stand-alone memory prompts.

The Learning Approach

Getting better at anything benefits from an open mind that welcomes challenge and reflective decision making. Talk-for-learning is an effective approach for QuBuild when there is a culture of collaboration that encourages and celebrates child-led scientific question-asking.

Teacher talk examples of the QuBuild Process in context

The authors have spent many, many hours speaking and working with primary science teachers. They have researched practice in science enquiry and question-asking and have pondered on the experiences of teachers' trials and tribulations. The book starts and closes with teacher voices. Listen to the voices of four teachers in different contexts and reflect on how their experiences are similar or different to yours.

Teacher A

My context

I am an experienced primary school teacher with a science background. My own subject knowledge for the chemistry concepts is good. I believe in child-led learning and encourage children to find out answers through enquiry.

My need for a solution

For many years I have set the children the task of investigating sugar in tea. This is fairly common in many schemes of work. The children change one variable and measure another in a typical comparative test enquiry approach. I support the children to ask questions with two variables in, as I think this supports them to write better conclusions. However, at the end of this task I often feel frustrated with the outcomes. I find that too many children use key terms such as melting and dissolving incorrectly and interchangeably in their conclusions and discussion. I am concerned as children clearly remain confused and whilst their learning of variables is good, the substantive knowledge and their preconceptions about core concepts have not been challenged sufficiently.

My lightbulb moment!

Prior to learning about the QuBuild Process, my solution to the problem was to spend the next lesson with me acting as 'teacher expert' - I often found myself telling the children the things that they got wrong, and in some cases, adding in the corrections. This approach is highly teacher-led and the children responded to me by doing as I requested. My issue here is that they had little ownership of the new learning and I don't fully believe that the confusion wouldn't arise again in the future.

QuBuild has enabled me to value all types of questions and to understand that there are different questions for different reasons - I have embraced the thinking that it's more to do with, 'the right question at the right time for the right reason'. I no longer think that closed questions are worse than others, they are just different and fit for a different purpose.

My practice has now changed. I have now focused initially on Questing-producing. The children work together to produce as many questions as possible with a range of different question-stem starters. This is fairly straightforward, and my role was to make sure everyone contributed. Next, we sort the questions into different groups using the definitions of the Question-handling tools. This stage is harder for the children, so I have simplified it by asking them to find all the closed questions and place them into one pile and place all the other questions into a different pile for use later.

The learning moment for me and the children has been the Question-improving stage. Having spotted the simple tactic of 'tweaking questions', the children are now better at picking out closed questions and improving them to be 'better' closed questions. This was new yet didn't take long to master. The benefit has been that children were now curious about the facts that I wanted them to know about. I feel much more confident that the key concepts that I know are important and need to be understood are developed in order to plan a double-variable scientific question.

After tweaking the closed questions, Question-choosing now took place using the traditional sticky-note method and planning boards. Quality conversations about evidence and trust of evidence have started to happen. This means the children are more authentically working like scientists even before the equipment comes out for the task.

By spending more focused time on the QuBuild Process, the learning was deeper and more memorable to the children. Their ideas about melting and dissolving had changed, their understanding of disciplinary knowledge of quality of evidence had developed and their ability to draw conclusions with cause and effect relationships was secure.

Exemplification School A

From CLOSED questions	Tweaked to improved CLOSED questions
Does sugar melt?	Does **brown** sugar melt? Do **all types** of sugar melt at the same temperature? Does **lumpy** sugar melt easier than non-lumpy sugar?
Does sugar dissolve?	Does sugar dissolve **in water**? Do **all types of** sugar dissolve in water? Does **caster** sugar dissolve in cold water? Does sugar dissolve **in milk**?

From CLOSED Questions	Switched to improved alternative questions
Does brown sugar melt?	What temperature does brown sugar melt at? **Fact finding**
Do **all types of** sugar dissolve in water?	What will we see if the different types of sugar dissolve? **Empirical question**

The children said:
"If we have a fact-finding question we can gather the evidence using secondary sources. It's likely someone else will have done this or something like it before. The Quality Checker says we should try two secondary sources and compare them".

The children found:
Two websites that gave different answers, one said 186 degrees Celsius and the other one 200 degrees Celsius. The children were fascinated at the evidence and wondered who to trust.

The children took action:
Finding the thermometers in the school science cupboard they noticed that they would not help them as they only read to 110 degrees Celsius. They decided to find out who might be an expert that they could trust.

The children started to understand:
That the school equipment is good for dissolving but not for melting. Dissolving and melting are not the same thing.

Teacher B

My context

I have been teaching science for many years and always seek ideas for practical work that relate to the topic I am teaching. In my lessons we try to move on to the practical set up as quickly as possible as the children enjoy it, and I don't like to run out of time when the lesson is so short. I suppose when I think about it my approach to science is to try to get on with the good stuff as quickly as possible.

My need for a solution

When I teach plants, I include the practical activity of food colouring and celery. I know some schemes also use a carnation flower, yet celery is cheap enough that I can make the group sizes small enough that all children can be involved. I know that having a question to explore is good practice and so I display the lesson title as a question, typically the question I use is: Does the celery turn red with the dye? However, when I review the children's learning from this task, I find I am spending time marking very short conclusions. In fact, I realise that the answer to the question is simply 'Yes', and that I am having to add many prompts to encourage children to appreciate the link between the activity and the learning concepts about plants. I worry whether all the children have actually learnt that red food dye makes celery red, whereas I really want them to know about the parts and functions of the plant.

My lightbulb moment!

By using the QuBuild Process children took more ownership over the enquiry questions. The Question-producing step had different tools that were easy for the children to use. I was concerned that if I took time out of the lesson to build questions there would not be enough time to do the practical activity so I made the question producing just 15 minutes in the afternoon of the day prior to the practical science lesson.

I started the 'food dye and celery lesson' by modelling the steps of Question-handling and Question-improving and choosing. I explained that I wanted the children to be gathering their own data so we did a class shuffle and sort of all the questions that had been created by the children. Together we question-handled by separating from the collection the questions that were closed or fact finding. The questions that were remaining went forward for Question-improving. Using the 'Variable Spotter' and demonstrating the actions on the board, it was clear which questions needed tweaks to create a cause-and-effect question.

The enquiry type that I wanted the children to experience was observation over time. We talked about the last time that the observation over time type was used. The children noticed that the graphs all had time as the dependent variable (well they actually said it is always time on the bottom). We sorted through the questions to identify those that might help find out more about the movement of water by gathering information with time as the thing that we measured. The scientific **question chosen** became: *Does the number of leaves on the celery stalk affect the time taken for red food dye to travel up the stalk?*

My lightbulb moment was that there is a link between the type of enquiry and the question type. I had actually never thought about this link before. For the children it was evident that their interest, curiosity, and confidence to work scientifically was enriched when I overheard the children the next week chatting about what they had found out about the movement of water through plants by doing their own experiments at home and reading more about plants. One child said that the milk wasn't as good because they couldn't see when or if it had reached the leaves, and another one was heard wondering if blue food colouring might make carrots purple and that grandma said that carrots hadn't always been orange.

Exemplification School B

The QuBuild Process	The children's responses
Question-producing **The Question Spinner**	**Children created lots of questions.** • Should plants always be watered through the soil? • Can plants grow if they have no roots? • Will the water go through the surface of the leaves? • Do all plants have stems? • What other liquids might travel through a plant stem? • Why do plants need water? • Which plants die first if they don't have any water? • What is inside the stem?
Question-handling **Question Shuffle and Question Catalogue**	**Children sorted their questions and talked about them.** Fact-finding and or closed: • What is inside the stem? • Why do plants need water? • Do all plants have stems? • Will the water go through the surface of the leaves? Not fact-finding and or open: • Which plants die first if they don't have any water? • What other liquids might travel through a plant stem? • Will the water go through the surface of the leaves? • What is the best way to water a plant? • Does the age of the celery stick affect it?
Question-improving **Variable Highlight and Tweak Tactics**	**Children changed and adapted questions together.** • Which of the different plants in our classroom wilt first if they don't have water? • Which liquids travel fastest through the stalk? • Do some plants have faster movement of water from the roots to the leaves? • Does celery with more leaves work better than celery with less leaves? • Which grows best: a plant watered on the saucer or a plant sprayed on the leaves? • How fast does red food colouring move along the stem? • Does food dye move quicker in older celery than freshly picked? • What is inside the stem?
Question-choosing **Keywords Science – Observation over time**	**Children agreed that the question is fit for purpose.** The question will lead to evidence that can be collected. The question is a good scientific question. • Does the number of leaves on the celery stalk affect the time taken for red food dye to travel up the stalk?

Teacher C

My context

I am the subject leader for science in a two-form entry primary school. The area for development and improvement across our school was progression in working scientifically. Science has had a high profile for many years and there is whole-school engagement in science week every March. A special science week provided a time for every year group to use the same activity of exploring the relationship between hand size and how many sweets they could pick up. This was presented in an open way so that children could make their own choices, find out results and draw conclusions from their evidence.

My need for a solution

The range of work produced by the children stimulated much staff discussion and was certainly interesting. Some of the questions included in the work samples were:

- Can taller people hold more things?
- Does the oldest person have the biggest hand?
- Do bigger hands pick up more sweets?
- Will using the square sweets be easier than the round ones?

My lightbulb moment!

By being aware of the QuBuild Process I realised the link between the quality of the question and the conclusion that could be drawn. Indeed I realised that whilst as adults we were expecting a conclusion that described and explained the relationship or pattern found, if the question asked was identified to be a closed question then the conclusion would be only one word – 'Yes' or 'No'. I led a staff meeting on **Question-handling** and **Question-improving** so that there is now a better understanding of how to help the children create cause-effect questions that are open. As a result, the conclusions and scientific thinking from the children have improved.

- How does the size of the handspan affect the number of sweets picked up?
- How does the shape of the sweet affect the number of sweets picked up?
- How does the age of the person grabbing the sweets affect how many sweets they picked up?
- What happens to the number of sweets if the grab is from above or from underneath?

Teacher D

My context

I have been teaching 9-10 year olds for many years and I embrace new ideas as they come to my attention. I really thought about the scientific question-asking key messages and have taught the activity resources in my class. I remember when the equipment children could access for help was simply pencils, rulers and scissors. Now I love that there are so many more hands on easy to use tools for my children to use.

My need for a solution

In a class of thirty children it is really difficult to help every child with every need that they have at the moment they seek help. I only have a teaching assistant in the morning and miss the extra pair of hands when doing science in the afternoon.

My lightbulb moment!

I absolutely love my new **QuBuild Kit-box**. In the centre of every table I have put a tray of the QuBuild tools so children can simply reach, grab and use them without asking me to run around fetching things. I can focus on those that need my help and observe with delight those that are active, proficient and independent scientific question askers. Furthermore the tools from the QuBuild Kit-box are beginning to be used in other subjects now too.

In the **QuBuild Kit-box** children have:

★ One Question Spinner

★ One set of brightly coloured 'Wonder Bubbles' cards

★ One Question Teller

★ One Question Sort flashcards

★ Two highlighter pens of different colours

★ Sticky-notes

★ Question Checklist A4 clipboard

★ QuBuild Question Keywords

★ QuBuild Science Keywords

★ One copy of each of the QuBuild visual icons printed on A5 card

References and further reading

Bianchi, L., Whittaker, C. & Poole, A. (2023) *Being Focussed: Monitoring the 10 Key Issues to Improve Children's Learning Experiences in Primary Science.* The University of Manchester and The Ogden Trust.

Bonsall, A. & Bianchi, L. (2019) *Children's scientific question-asking – an initial scoping of academic literature.* Journal of Emergent Science, Winter 2019/20; Vol 18, 29-34, The Association for Science Education.

Çalik, M., Coll, R.K. (2012) *Investigating Socioscientific Issues via Scientific Habits of Mind: Development and validation of the Scientific Habits of Mind Survey.* International Journal of Science Education. Vol 34: 1909–1930.

Chouinard, M. M. (2007) *Children's questions: a mechanism for cognitive development.* Monogr Soc Res Child Dev.;72(1):vii-ix, 1-112; discussion 113-26. DOI: 10.1111/j.1540-5834.2007.00412.x. PMID: 17394580.

Costa, J., Caldeira, H., Gallastegui, J. R. and Otero. (2000) *An Analysis of Question Asking on Scientific Texts Explaining Natural Phenomena.* Journal of Research in Science Teaching. Vol. 37, No. 6, PP. 602- 614

Great Science Share for Schools. Visit www.greatscienceshare.org. The University of Manchester.

Harlen, W. 2000. *The Teaching of Science in Primary Schools.* London: David Fulton.

Holt, J. (1971) *How Children Learn* (London, Penguin).

OFSTED (2021). *Research Review Series: Science.* Access at www.gov.uk/government/publications/research-review-series-science

Rothstein, D. & Santana, L. (2015) *Make Just One Change: Teach Students to Ask Their Own Questions.* Harvard Education Press.

TAPS (2020) *Focused Assessment: Teacher Assessment in Primary Science (TAPS) support for Working Scientifically.* Bristol: Primary Science Teaching Trust. Accessed at: https://pstt.org.uk/unique-resources/taps [pstt.org.uk]

The Right Question Institute, Access at: https://rightquestion.org/what-is-the-qft/

The University of Manchester (2021) *QuSmart Final Project Report.* Presented to the Primary Science Teaching Trust. Project team: L Bianchi, A Bonsall, B Turford, C Whittaker

Turner, J. 2012. *"It's Not Fair."* Primary Science. Association for Science Education. Vol 121: 30–33.

University of Berkeley. *Understanding Science and how it really works.* Access at: https://undsci.berkeley.edu/about.php

Watson, R., Goldsworthy, A. & Wood Robinson, V. (1998, 2000) *AKSIS-Science Investigations.* Education in Science, n187. p14-15 Apr 2000.

Acknowledgements

This book draws on contributions from several sources. We thank them for their inspiration and advice.

The AKSIS team and the Right Question Institute, both who have inspired us and made us ask a lot of questions about scientific questions. We take forward their work in championing the need for asking better scientific questions within a primary school context.

The University of Manchester's Science & Engineering Education Research and Innovation Hub team, who collaborated on a range of projects that informed thinking towards this book.

The Primary Science Teaching Trust and Comino Foundation for funding to support a range of research and innovation projects over many years. Insights gained over this time have undoubtedly contributed to the awareness of the issues in schools related to scientific question-asking. In particular, we acknowledge the support and insights gained from the QuSmart project (Bonsall & Bianchi 2019; The University of Manchester 2021).

Trusted colleagues who provided valuable support and expert challenge during the development of this book with particular note to Anne Goldsworthy, Dr Sarah Earle, Dr Jenny Watson and Andie Hughes.

Tina and Lynne wish to acknowledge the support given by family and friends over the time of writing this book. In particular, many hours checking for full stops, querying commas and generally supporting their professional time to discuss and develop this into something to be really proud of!

Science & Engineering Education Research and Innovation Hub (SEERIH) is a nationally recognised centre of science and engineering education. Based in The University of Manchester its mission is to develop and engage teachers in innovative, research-informed continuous professional development programmes to ensure high-quality learning outcomes for young people. For further information visit **www.seerih.manchester.ac.uk**

Design & artwork by David Webb. Cover font design by Nils Cordes.

About the authors

Professor Lynne Bianchi, Director of the Science & Engineering Education Research and Innovation Hub, Vice Dean for Social Responsibility, Equality, Diversity, Inclusion and Accessibility (The University of Manchester).

Lynne is a specialist in curriculum and professional development, innovation and research in primary science and engineering education working. Having qualified as a teacher, she worked in schools in Greater Manchester before achieving her PhD in Science Education. She developed her work in curriculum and teacher development at the Centre for Science Education at Sheffield Hallam University for 14 years.

In 2014, she launched SEERIH at The University of Manchester, developing it as a leading UK centre of expertise, with a vision to improve the attainment of children in schools in areas of high socio-economic deprivation. Lynne is a national leader in primary science and engineering education, working as an advisor and innovator with a wide range of Learned Societies including the Royal Academy of Engineering, Royal Society and Institute of Physics. She is passionate about enhancing inclusive experiences through large scale campaigns, in particular the flagship Great Science Share for Schools campaign. In 2022, Lynne became a Vice Dean in the Faculty of Science & Engineering further enhancing her ambition to make a difference by championing and developing the very best educational opportunities for young people in STEM education.

Christina Whittaker, CSciTeach, Professional Development Champion (SEERIH), Senior Regional hub-leader PSQM, Strategic lead Science Across the City (Stoke-on-Trent).

Tina has over 25 years experience in contributing to and challenging school improvement through reflective dialogue. As such she has spent most of her career asking questions and many would say the 'hard' questions. Tina is particularly passionate about quality science education for all children and building effective leadership at all levels. The ability to ask good questions connects the discipline of science and the role of being a leader. The skill of being able to develop and ask better questions results in better informed decision making regardless of the context. Tina has been in the role of senior PSQM hub leader for almost 10 years, supporting schools to plan for impact through the essential 'so what' question. In 2019 she took her hard questions to Stoke-on-Trent city leaders making the case for primary science to be included in the strategy for social mobility with parity alongside English and Mathematics and so establishing the initiative Science Across the City (SATC).